ROBERT S. DEVINE

THE SUSTAINABLE ECONOMY

Robert S. Devine has been writing about the environment, politics, natural history, and the outdoors for decades. His books include *Bush Versus the Environment, Alien Invasion: America's Battle with Non-Native Animals and Plants,* and more than ten books he has authored or coauthored for the National Geographic Society. His articles have appeared in dozens of publications, including *The Atlantic, Audubon,* and the *Los Angeles Times.* He lives on planet Earth, which he hopes will be a good place to live for all generations to come.

ALSO BY ROBERT S. DEVINE

Bush Versus the Environment
*Alien Invasion: America's Battle with Non-Native Animals
and Plants*

THE SUSTAINABLE ECONOMY

THE SUSTAINABLE ECONOMY

THE HIDDEN COSTS OF CLIMATE CHANGE AND
THE PATH TO A PROSPEROUS FUTURE

ROBERT S. DEVINE

ANCHOR BOOKS
A DIVISION OF PENGUIN RANDOM HOUSE LLC
NEW YORK

AN ANCHOR BOOKS ORIGINAL, OCTOBER 2020

Library of Congress Cataloging-in-Publication Data
Names: Devine, Bob, 1951– author.
Title: The sustainable economy : the hidden costs of climate change
and the path to a prosperous future / Robert S. Devine.
Description: New York : Anchor Books, 2020. | Includes
bibliographical references and index.
Identifiers: LCCN 2019059208 |
ISBN 9780307277176 (trade paperback)
Subjects: LCSH: Economic development—United States—History—
21st century. | Environmental policy—United States. | Climatic
changes—Economic aspects—United States. | Climatic changes—
United States.
Classification: LCC HC106.84 .D48 2020 | DDC 333.70973—dc23
LC record available at https://lccn.loc.gov/2019059208

Anchor Books Trade Paperback ISBN: 978-0-307-27717-6
eBook ISBN: 978-0-593-31117-2

Author photograph © Trudy Ewing
Book design by Nicholas Alguire

www.anchorbooks.com

Printed in the United States of America
10 9 8 7 6 5 4 3 2 1

To all the dedicated people who are working so hard to create a sustainable, just, and prosperous society. Given their efforts, I like our chances.

CONTENTS

THE SUSTAINABLE ECONOMY

INTRODUCTION

In this book we'll confront two scary subjects: climate change and economics. I'm not sure which is scarier. The former involves an existential threat to civilization and the latter involves a lot of math. But not to worry. Importantly, I'll spare us from the math. More importantly, I'll explore some economic ideas that can spare us from much of the impact of climate change. Most importantly, these same ideas can help us reshape our economy to create a sustainable, just, and prosperous society that better aligns with our values.

Does this sound too good to be true? Well, note the presence of the word "can" in the last two sentences of the above paragraph. Not "will" but "can." Yes, we *can* blunt the effects of global warming by making changes to our economic system that also *can* improve our society. But this *will* happen only if we, the people, make it happen. And I'm not talking about our personal behavior, such as choosing to buy energy-efficient appliances, though such individual actions are important. I'm talking about enough of us acting together to create systemic change, such that we have an economy that produces only energy-efficient appliances.

To begin this book's journey, come with me for a moment to Central California. One October day a few years back I was driving along a back road in the San Joaquin Valley. Autumn be damned, the unblinking sun had propelled temperatures uncomfortably close to 100°F. With the AC cranked high, I motored along a dusty gravel road past some droopy cotton fields and almond groves. Surprisingly, I also encountered acre upon acre of cracked, hard-baked dirt where crops had not been planted that year—surprising because this valley arguably ranks as the most productive farming region in the world. But even the valley's agricultural prowess and unconscionable amounts of chemical inputs couldn't overcome the lack of water.

My visit occurred during the drought that desiccated California for many years, starting in 2011. It was so brutal that the drought map of the state showed only a few areas lucky enough to be suffering merely "severe" drought (the third-driest condition) or "extreme" drought (the second-driest condition). The vast majority of the color-coded map was darkened by the brick red of "exceptional" drought, the driest condition possible. Entire towns ran out of drinking water. Wild animals starved to death by the thousands. The drop in crop production forced many residents to move in search of work. More than one hundred million trees died. Wildfires blazed through dried-out forests and parched towns. Fish struggled to survive in the warm waters of the low-flowing rivers. Hungry people crowded around mobile food banks set up in sweltering parking lots. Incidents of domestic violence increased in hard-hit farming communities. Poor farm workers sank deeper into poverty. When drought shrivels the land, it also shrivels lives.

Droughts have been part of life on earth since before humans existed, so it's understandable that people may think of California's destructive dry spell as a purely natural disaster.

But it was not. Some of the blame belongs to the unnatural disaster of climate change, which is ramping up droughts in many parts of the nation and the world. And the ultimate blame for climate change belongs to the unnatural entity that is our economic system.

Global warming is a market failure. On the surface, climate change is an environmental crisis that results from overloading the atmosphere with greenhouse gases, but in large part fundamental shortcomings of our market generate the excessive emission of those gases. In a landmark report for the UK government on the economics of climate change, Nicholas Stern, a prominent scholar and former chief economist at the World Bank, writes, "Climate change presents a unique challenge for economics. It is the greatest and widest-ranging market failure ever seen." I agree with him except for one word in his first sentence: "unique." Yes, climate change presents a huge challenge for economics but, sadly, not a unique one.

Though it sounds like a generic phrase, "market failure" actually is a technical term in economics. It does not refer to insider trading, corporate fraud, real estate scams, or any of the other business misdeeds that grab media headlines. Definitions vary, but essentially a market failure occurs when people experience a net welfare loss because the marketplace is not optimally allocating goods and services. Broadly speaking, fraud and scams stem from bad behavior by individuals, but market failures stem from flaws deep in the DNA of our economic system. It's not that the market *doesn't* prevent or correct market failures; it's that the market *can't*.

To guide us through the maze of the market failure that is global warming, this book will delve into the social cost of carbon (SCC). Dozens of governments around the world employ some form of this economic tool in an effort to help reduce carbon dioxide emissions, but in this book we will come to grips with America's federal SCC.

In part, I've chosen to focus on the SCC because it's important in its own right. However, my main reason for focusing on the SCC is that it provides a revealing lens through which to examine the economics of climate change, the workings of markets, and some of the key ways in which we need to improve our economic system. Unfortunately, what we see when we peer through the lens of the SCC is unnerving: our current version of a market is incapable of producing a sustainable, just, and prosperous economy. We need to make a transition to what I call "sustainability economics."

Sustainability economics is not a formal academic field. It's a term I use to consolidate the ideas and research I've marshaled for this book, which come from people in a variety of actual fields, such as ecological economics and environmental economics, as well as from activists, policy makers, and other non-scholars.

Now, let's meet the social cost of carbon. The SCC is an estimate of the value of the impact on society over time from the emission of one metric ton of carbon dioxide into the atmosphere. This estimate is primarily used in regulatory procedures to help decision-makers better understand the costs and benefits of proposed rules about everything from the energy efficiency of toasters to the carbon pollution from power plants. The influence of the SCC can reach well beyond proposed regulations, too. For example, it might play a central role in setting a price for carbon emissions.

The perspective underlying the SCC could prove decisive in the debate over climate action, especially considering that the recent surge of concern about global warming is eroding climate denial. Increasingly, fossil fuel boosters are retreating to their second line of defense: the claim that the costs to society of significantly curbing carbon emissions would outweigh the benefits. However, the claim that we can't afford to wean our economy from fossil fuels is dramatically inaccurate. In reality, we can't afford not to.

If you've been hearing about the rapid progress of clean energy sources like solar and wind, you might be tempted to think that we won't need the SCC or any other climate policies because market forces are bringing about the demise of fossil fuels. Please resist that temptation. Yes, the rise of clean energy is heartening, and solar and wind are becoming competitive with fossil fuels in some places. But no, the shift is not happening nearly fast enough to stave off damaging and perhaps cataclysmic climate change. Besides, we owe much of clean energy's rise to government action, not just to the market. Likewise, we shouldn't count on miracles from geoengineering—the large-scale manipulation of the environment—because its methods are too unproven. Though innovative tech will help immensely, it's unlikely to take care of climate change on its own, especially not in the time frame science recommends.

To grasp the inadequacy of current efforts, just scan the headlines from the last couple of years. Study after study and report after report have underscored the fact that what we're doing isn't working well enough. In 2018, the Intergovernmental Panel on Climate Change (IPCC) sent shivers down humanity's collective spine by declaring that, for starters, we have to massively pick up the pace and reduce emissions by about 45 percent by 2030 in order to stand a good chance of avoiding grievous harm. Additional spine-shivering IPCC reports followed in 2019. The overall message can be summed up by this opening line from an IPCC press release: "Limiting global warming to 1.5°C would require rapid, far-reaching, and unprecedented changes in all aspects of society."

However, we're not exactly exploding out of the starting blocks when it comes to picking up the pace, as we learned in a report from the Global Carbon Project: global releases of carbon dioxide *rose* 2.7 percent in 2018 to an all-time high. The news doesn't get any better in the UN Environment Emissions Gap Report, which concludes that the emission-reduction goals of the heralded Paris climate agreement fall far short of the mark. To hit the temperature targets now endorsed by most climate scientists, nations would have to elevate their ambitions well beyond Paris levels—and currently most nations aren't even on track to meet their Paris commitments.

So, how is the United States doing? Before I answer, it's important to note that over the years many people have recommended that the global community reduce its greenhouse gas emissions 80 percent by 2050, though recent research indicates that target is too weak. But if 80 percent is too weak, I don't know what the adjective is for where American emissions are headed. A 2020 report from the U.S. Energy Information Administration found that on our current trajectory our nation's emissions will fall by only 4 percent between 2019 and 2050.

Some fossil fuel supporters try to brush off such frightening reports by portraying climate change as some remote process that, even if it's real, will exert little influence on our daily lives—perhaps a problem for polar bears but not for people. Consider a statement Mitt Romney made in 2012, in his speech accepting the nomination to be the Republican candidate for president. In an attempt to paint his opponent, Barack Obama, as out of touch with people's needs, Romney ridiculed Obama's efforts to address climate change: "President Obama promised to slow the rise of the oceans and to heal the planet. My promise is to help you and your family." As if reducing present and future droughts, floods, wildfires, diseases, wars, financial crises, heat waves, hurricanes, mass migrations, and sea level rise would not help you and your family in your everyday lives.

Allow me to hurl one more big international report at you because it broaches a crucial matter. The World Economic Forum's Global Risks Report for 2019 evaluated the risks associated with food shortages, failing governments, weapons of mass destruction, and dozens of other world problems. The report states that "over a ten-year horizon, extreme weather and climate-change policy failures are seen as the gravest threats." But that statement, though crucial, is not the particular crucial matter I want to underscore. I want to highlight the implication of the report's assertion that "of all risks, it is in relation to the environment that the world is most clearly sleepwalking into catastrophe."

The environment, it says, not climate change alone. Though climate change is the single most pressing environmental problem and it magnifies many other environmental problems, it is far from being the only pressing environmental problem. The forum finds in its report that three of the top five risks in terms of likelihood and four of the top five in terms of impact are environmental, with biodiversity loss and the subsequent ecosystem collapses ranking nearly as high as climate change. This reveals the crucial matter I want to stress—that climate change is by no means the only threat spawned by our current market system. Even if some basement-dwelling genius invents an atmospheric Roomba that immediately vacuums up all greenhouse gases and ends the danger of global warming, society would still be desperately in need of sustainability economics to deal with many other accelerating environmental crises.

For the most part we currently take a piecemeal approach to sustainability problems. We labor to restrict harmful chemicals, prevent species extinctions, stop sewage systems from overflowing into our rivers, protect marginalized communities from toxic dumping, and save critical wetlands from being paved over. Such endeavors are indeed vital, but they are stopgap measures that don't reach the core question: Why does our economic system enable and even encourage such harms? Running around trying to mop up all the water spewing from a burst pipe is exhausting and relatively ineffective. We need to fix the pipe.

Embracing sustainability economics would not mean abandoning our present market system, which broadly can be termed neoclassical economics. In a lot of circumstances it works well, and many of its principles are integral to sustainability economics. But we do need some basic changes. Like most nations, the United States has a mixed economy, meaning a blend of the private and public sectors. Especially during the last several decades, our mix has veered dangerously toward the private, to the detriment of public goods like the environment. A sustainable economy would involve more values-driven, citizen-supported public planning and less abdicating of decisions to the insentient workings of the market. Unfortunately, "plan" is a four-letter word in the laissez-faire lexicon and stirs up fierce resistance.

Kristen Sheeran, an economist and energy and climate change policy adviser to Oregon's governor, thinks this resistance indicates that adherents of the orthodoxy see the economics of climate change as an existential threat. And she thinks they're right. "Climate change really is the litmus test for neoclassical economics," said Sheeran. "This is clearly an issue where the standard tool kit of neoclassical economics doesn't apply. And what does that say then for neoclassical economics if it is inherently ill-suited to dealing with the most fundamental question of our time, a true civilization challenge?"

The ubiquitous influence of the economy in environmental issues provided the impetus for this book. Over the years, I've written about declining salmon runs, ranchers beset by invasive species, ailing kelp forests, the health effects of air pollution, excessive logging, biodiversity loss, ocean dead zones, vanishing wetlands, wildfires, the harm dams inflict on fisheries, and an assortment of other environmental topics. I was a little slow on the uptake, but after I'd done scores of articles and books, I finally realized that I should heed the famous saying "It's the economy, stupid." From sea otters to sea level rise, almost everything I was writing about circled back to the economy. For the most part we don't have environmental problems; we have economic problems.

Curing these problems will require systemic change, though that won't be a cure-all. We'll still have some greedy individuals who will not flinch at inflicting social costs on the rest of us as they grasp for riches and power. Especially damaging are the "predators," a term I borrowed from economist James Galbraith's 2008 book, *The Predator State*. This influential subset of greedy people and their political enablers loudly proclaim their opposition to government-instituted systemic reforms that decrease their profits while they are quietly lobbying for government favors that increase their profits. Deregulation makes up only half of the predator diet. Their other food group consists of subsidies. Predators chase down subsidies the way cheetahs pursue gazelles. Their appetite for government meat lays bare the hypocrisy of their complaints about government intervention in the market. Small wonder so many people feel cynical about government—that's just how the predators like it.

Though predators and other grasping individuals certainly do a lot of harm, this book gives them only minimal attention and instead focuses on the economy's many structural weaknesses and the changes needed to strengthen that structure. Besides, sustainability economics will better contain such predators than our current system, which can conceal and even promote such predatory behavior.

More important, sustainability economics will unobtrusively guide the rest of us, the vast majority of people who don't want to inadvertently harm other people and the planet. In today's system many of the actions we take and purchases we make impose social costs, usually costs that are so distant in time or space and so obscured by complexity that we're unaware of them. A sustainability approach will better align the economy with our values so that our transactions automatically take into account the associated social costs. This would relieve us individual consumers of the impossible burden of trying to comprehend the full social costs of every shirt, phone, or banana we buy and futilely struggling to incorporate that knowledge into every consumer decision we make. Likewise, a sustainability economy would relieve well-meaning producers of such burdens.

All things considered, embracing sustainability economics wouldn't mean lowering our standard of living. On the contrary, it would increase our quality of life. Sure, there's some stuff we would have to sacrifice. People would probably have to stop buying new phones every year and build fewer three-thousand-square-foot houses, but we'd still have plenty of stuff. Besides, the word "sacrifice" often connotes giving up something for nothing, but sustainability economics is about trade-offs and priorities. Yes, we would have to give up driving emission-spewing vehicles, but that doesn't mean we would have to sacrifice mobility. We could still get around, and get around more efficiently, by shifting to a mix of clean vehicles, mass transit, bike paths, and walkable cities. Of course, this shift means we wouldn't be able to spend as much quality time sitting in traffic jams or spend as much money getting treated for pulmonary diseases, but I'm confident we could learn to live with that sacrifice. And we'd enjoy the vital benefits—such as a livable climate—that necessitated our so-called sacrifices in the first place.

We need to keep reminding ourselves that stuff is only one element of a good life. Along with ample material wealth, a sustainable economy would provide us with other, often more valuable elements, such as improved health, better jobs, safer communities, and more free time. Besides, it's not as if nature is offering us the option to shun sustainability economics and blithely continue amassing ever more stuff, at least not for long. If we blunder onwards with our current market system, the biophysical foundation on which our economy rests will continue degrading until eventually our heedless binge will sputter to a stop. Then we'll have less stuff *and* we'll miss out on the non-stuff benefits of a sustainable economy.

According to the tradition of book introductions, by now I should have defined "sustainability economics," preferably using plenty of weighty words like "synergistic" and "optimize." But I haven't and I won't. A sense of the term's meaning will come into focus only over the course of the whole book, and, one hopes, without an appearance by "synergistic" or "optimize." For now I'll just mention one key point: sustainability economics flows from the understanding that our economy—and, for that matter, human civilization—depends entirely on the environment. The digital realm notwithstanding, we cannot live apart from soil, calcium, grass, nitrogen, bees, chlorophyll, slime mold, and all the rest of the biophysical world. This seems obvious, yet our current market system largely overlooks this simple but profound truth. Sustainability economics promotes the pursuit of the "Three E's": equity, the economy, and the environment. It recognizes that the first two E's can happen only if we sustain the environment—the environment is the bedrock—but sustainability economics also recognizes that if we don't address equity and the economy, we won't be *able* to sustain a livable environment. Sustainability economics is grounded in the knowledge that equity, the economy, and the environment are inextricably linked.

I'll also forgo defining "sustainability." But again, I do want to mention one key idea: contrary to the conventional framing, sustainability is not about saving the planet. Even if we humans continue to abuse the environment, we will not destroy the planet. We may debase the earth to the point that it is best suited for cockroaches, but the third rock from the sun will muddle through. Sustainability is about quality of life for people, about maintaining the environment in which we evolved and that allows us to thrive.

Some people—myself among them—would add that we *Homo sapiens* also have a moral responsibility to steward the rest of the earth's species. Other people think that nonhuman species don't rate much consideration apart from being a resource for people. Fortunately, this disagreement will be irrelevant if we achieve sustainability; fostering a robust diversity of other species is integral to the health of our environment, so actions that help sustain life forms from aardvarks to zooplankton are also essential to sustaining us humans.

While we're on the subject of subjects I don't cover in this book, note that even though tackling the SCC naturally involves abundant discussion of climate change, this book does not engage in the debate regarding the reality of climate change. Not only is that debate not my subject, but there is no serious debate. Decades of extensive scientific study have firmly established that climate change is real and primarily caused by human actions. In fact, recent research indicates that climate change is happening faster and will hit us harder than most scientists thought just a few years ago (unless we take swift and decisive action, like, say, adjusting some fundamentals of our economy).

The uncertainty about the science of climate change—and there is plenty of uncertainty—lies in the details about its timing and impacts. How well will animals and plants adapt? What is the probability that Antarctica will experience a major meltdown in this century? How often will climate-juiced hurricanes thrash our communities? This book will probe some such uncertainties because they're integral to understanding the SCC and the economic risks of a chaotic climate. But as to the reality of climate change, it rests firmly in the settled-fact file along with the knowledge that the earth is not flat.

Let's move on to what I do cover in this book. The first seven chapters focus on the inherent weaknesses of a market economy. Ideas about how to strengthen these weaknesses crop up throughout the book, but the full discussion of how to achieve better living through sustainability economics comes in later chapters.

The first three chapters dig into the social cost of carbon and some of the central assumptions that produce it, assumptions that reveal why markets by their very nature can't provide many of the goods and services most dear to us, such as a livable climate. We'll examine issues such as the market's myopia regarding biophysical reality and our economy's structural biases that favor private goods over public goods. Chapter four focuses on how risk is treated when we calculate the SCC, exposing ways in which neoclassical economics underestimates some of the low-probability, high-impact dangers of climate change. Chapters five through seven wrestle with the equity implications of our market's inability to handle climate change and other environmental problems. Chapter five looks at intergenerational equity by following a group of dedicated kids and youth who have taken the U.S. government's climate change policies to court. Chapter six shows how a standard financial practice called "discounting" epitomizes an economic philosophy that can eviscerate the SCC, undermine climate action, and shortchange the future. Chapter seven shifts to present-day inequities that stem from climate change and dissects some of the neoclassical tenets that enable those inequities.

The three middle chapters challenge several of the market economy's favorite answers to the problems covered in the early chapters. Chapter eight considers the contentious efforts to account for the full costs of climate change by establishing a carbon price—efforts that reveal the market's limited ability to carry out such accounting whether for global warming or other environmental and social harms. Chapter nine looks under the hood of orthodox economic growth—laissez-faire's all-purpose solution—and finds that some of the cylinders in this internal-combustion clunker are misfiring. We'll learn why old-style growth won't make up for the harm caused by global warming and how misguided reliance on such growth can hamstring climate action. Chapter ten debunks two key market myths. One is that technological innovation can cure most if not all of our environmental ailments, including global warming. The second is that when society exhausts a resource, the market will always generate a substitute that is just as good or even better.

Though every chapter sheds light on ways to make our economy and society more sustainable, it is in the final two chapters that we'll pull everything together to illuminate the path to a sustainable, just, and prosperous future. Lots of creative people have lots of promising ideas, some of which are already being put into practice. We'll look at a few of those, but I won't be presenting long lists of concrete actions. Instead, I'll spend most of chapters eleven and twelve discussing a few overarching changes in attitude and perception that we must make in order to set the stage for those concrete actions. We'll contemplate the need for a more mature understanding of freedom. We'll judge the pivotal role human judgment must play in making decisions that we've been ceding to the market. We'll ponder the conundrum of individuals choosing to take collective action. We'll talk about ways in which we can create jobs that are more sustainable and more meaningful—and that pay better. These concluding chapters range widely and take us to surprising places.

The essential issue these last two chapters probe concerns the basic purpose of an economy: allocating limited resources—labor, capital, and natural resources—among competing ends. Laissez-faire disciples would say we shouldn't even be asking such questions because the invisible hand of supply and demand will determine optimal allocation for us. Sustainability economists would say we must ask such questions if we want to overcome the market's blind spots and have our economy put enough resources into enterprises that create the greatest well-being.

Ironically, neoclassical economics provides a concept that conveys the need for our economy to evolve beyond neoclassical economics: creative destruction. Neoclassical economists typically use this term when an outdated technology or business is shoved into the landfill of history by something new and better, as when computers ousted typewriters. But the term also can be applied to social innovation. Perhaps the market faithful will find it easier to accept sustainability economics if they think of it as the creative destruction and replacement of those aspects of the market that are outmoded.

It's not surprising that conventional economic thought needs a makeover, given that its framework originated well over two hundred years ago. If you had to choose a birth year for our current form of market economics, 1776 would be a likely candidate. The Industrial Revolution was beginning to gather steam, but there's a more specific reason to single out that precise year: it marks the publication of the Scottish economist Adam Smith's *The Wealth of Nations*, the seminal work of free market beliefs. Due to the explosive expansion of industry and the subsequent boom in global population from about eight hundred million in Smith's time to almost eight billion today, human civilization changed more during these last two and a half centuries than during all of human history before 1776, altering the world in ways Smith could not have imagined. Remember that Smith wrote *The Wealth of Nations* at a time when waterwheels powered mills, farmers used horses to plow fields, and many people still hunted for their dinners—with muskets.

Though the concepts of Smith and his ideological descendants revolutionized economics and still offer many valuable ideas, the time has come to update our economic system to fit the twenty-first century and the future. The physical frontier on earth is largely gone, but the social cost of carbon and sustainability economics are part of a cultural and intellectual frontier that we've only begun to explore. Think of this book as a scouting report from a promising and exciting new land.

1

THE PRICE IS WRONG

"Let's cut to the chase. What's the justification for assigning zero value to CO_2 reduction?"

The question from Michael Hawkins stopped Thomas Byron mid-sentence. This interruption might have seemed rude in other settings, but Hawkins was a judge on the U.S. Court of Appeals for the Ninth Circuit, and Byron was an attorney appearing before a panel composed of Hawkins and two other judges. With limited time for oral arguments, appeals court judges often cut in when they want to steer an attorney to what they consider the heart of the matter. Still, Hawkins's voice did carry a note of impatience.

These oral arguments occurred on May 14, 2007, in the case of *Center for Biological Diversity (CBD) v. National Highway Traffic Safety Administration (NHTSA)*. Joined by eleven states, the District of Columbia, the City of New York, and four public-interest organizations, CBD had challenged NHTSA's latest fuel economy rule for light trucks. CBD et al. thought the rule should require more miles per gallon from more types of vehicles. Although only two attorneys from each side argued the case, its importance drew some twenty other lawyers associated with the litigants to the ornate turn-of-the-nineteenth-century courtroom in San Francisco. Scores of interested onlookers filled the rest of the seats, among them then California attorney general Jerry Brown, soon to be elected as the state's governor.

CBD v. NHTSA featured many elements, including the high-profile battle over the loophole that allowed SUVs, minivans, and some pickup trucks to meet a lower fuel economy standard than cars, a loophole that largely accounted for the packed courtroom. But, surprisingly, the judges focused much of their scrutiny on an arcane dispute about the value of reducing carbon dioxide emissions and thereby reducing the impacts of climate change.

NHTSA asserted that the federal government's methods for calculating the benefits of spewing less CO_2 into the atmosphere lacked credibility. Consequently, in its cost-benefit analysis, NHTSA did not include any value for the reductions in CO_2 that would occur if the government required light trucks to go farther on a gallon of gas. In effect, NHTSA—and, by extension, the George W. Bush administration—had set the value of reducing CO_2 emissions at zero, a result that concerned Judge Hawkins and that led to his question to Byron: "What's the justification for assigning zero value to CO_2 reduction?"

"The agency did not assign zero value," Byron replied. "It concluded there was no way to assign a monetary value." According to Byron, NHTSA did acknowledge the benefits of lowering CO_2 emissions, but the agency couldn't include those benefits in its analysis "because the range of values [in the scholarly literature] is too imprecise to assign any particular monetary value."

"Did that range start at zero?" asked Hawkins.

"It started at $3 per ton of carbon, very close to zero," answered Byron. He did not mention that several studies put the value at about $50 a ton or that some estimates soared far higher.

"Your clients set it at zero," said Hawkins pointedly.

"I don't think it's fair to say that the agency chose zero as the value," said Byron. "That's not what the agency said in the final rule. To be sure, the agency didn't assign any value."

To which Hawkins replied, "That's like saying, 'Yes, we have no bananas.'"

Hawkins was not the only judge on the panel who found NHTSA's arguments lacking. The court unanimously ruled in favor of CBD, labeling the Bush administration's new fuel economy rule "arbitrary and capricious"—the standard legal phrase for "wrong." As its first reason for invalidating the rule the panel cited NHTSA's "failure to monetize the value of carbon emissions."

Years later I discussed the case with Sean Donahue, one of the two attorneys who had argued on the side of CBD back in 2007. He speculated that if the Bush administration had assigned even a paltry value to CO_2 emissions, such as $3 a ton, the court might have considered that adequate, given that courts typically defer to agency expertise regarding technical matters. "But zero is such a striking number," said Donahue.

Donahue noted that NHTSA's decision to go for zero fit with the George W. Bush administration's general approach to climate change: draw a hard line and thwart action whenever possible. "The Bush administration would have recognized that [assigning any value to reducing greenhouse gas emissions] would be a significant step, and they didn't want to take it." Declaring in court that a ton of CO_2 causes even three dollars' worth of measurable harm might have opened the door to a discussion the Bush administration didn't want to have, a discussion that might have led to cost-benefit analyses using a figure like $50 a ton. And the higher that dollar-per-ton estimate of the cost to society from carbon emissions went, the more likely a proposed regulation, such as the fuel economy rule, would be adopted despite opposition from industry. More broadly, the administration simply wanted to squelch any acknowledgment of the reality and impact of climate change.

Another reason NHTSA might have won the case if it hadn't taken such an extreme position is that their basic objection had some merit; the ways in which the costs of CO_2 emissions had been calculated were indeed "imprecise," as Byron put it. Analysis of the costs and benefits associated with rising CO_2 levels had been scarce and scattered prior to *CBD v. NHTSA*. It was this case, the first case in which a U.S. court required a federal agency to assess the monetary impacts of climate change as part of an environmental review, that in part led to the federal government's establishment, in 2009, of a working group to determine a social cost of carbon. However, as we'll see in subsequent chapters, despite the considerable progress made by that working group in 2009 and in later years, the argument over imprecision has continued, often at high decibel levels.

Though hardly a seismic moment that shook up American climate change policy, this case does make a point that could and should shake up not only climate change policy but also the economic landscape in general, a point that should steer society toward sustainability economics. Though its larger significance wasn't discussed, this point implicitly surfaced at times during the oral arguments. For example, in the course of questioning Donahue about the difficulty of assigning a dollar value to the emission of a ton of carbon, Judge Hawkins and Donahue had the following exchange:

"Your position," said Hawkins, "is that while there may be some variance in what experts say about monetizing this, it's not reasonable and might well be arbitrary and capricious to set it at zero."

To which Donahue replied, "We think that the one answer that can't be right is zero."

What should the number be instead of zero? What are the alternatives to using any SCC number at all? What do all these calculations and deliberations tell us about our economic system? The rest of the book engages with these and other profound questions the SCC raises, but first this chapter sets the stage by dealing with the truly seismic question embedded in Donahue's statement: Why can't zero be the right number?

LET MARKETS RULE US

The dim video from years ago shows a choir of about thirty college-age singers, women dressed in blue shirts and black skirts and men in white shirts and black pants. Arrayed against a dark curtain, the choir opens with the sopranos' solemn voices, soon joined by altos and the deeper tones of the young men. It sounds like a hymn, and in essence it is, but instead of glorifying the Almighty it celebrates the god of the free market. Lyrics include such poetic gems as "Let's use free markets as our tool," "Competition will make all the rules," and, near the end, "Let markets rule us."

This curious performance comes courtesy of the Milton Friedman Choir. Milton Friedman was a renowned economist and winner of the Nobel Prize for economics in 1976. When he died, in 2006, *The Economist* magazine wrote, "He was the most influential economist of the second half of the 20th century . . . possibly of all of it." Despite Friedman's stature, the Milton Friedman Choir's paean to the free market did not go platinum. However, his ideas, and the flood of like-minded ideas that soon followed, went platinum and then some. During the final quarter of the twentieth century, the belief that a minimally regulated market produces an optimal economy became orthodoxy in the United States and throughout much of the world. Not only did this vision of the omnipotence of the market come to dominate academic economics; it also became the majority view in business, government, and among the general public. Not everyone believed in this vision as purely as Friedman did, and some merely used it to justify their excessive profit-seeking, but the choir's exhortation to "let markets rule us" largely became the conventional wisdom.

The volume of media exalting the free market could fill a small library, but to get us started I'm going to describe the idealized basics in just three sentences, as follows: The market provides the greatest welfare to the greatest number of people by optimally allocating resources through the miracle of decentralized choices by the consumers and producers of goods and services. Though market theory bristles with inscrutable terminology and daunting equations, they're all branches and leaves that stem from a single trunk: supply and demand. Instead of commerce dictated by kings or commissars, the commerce in the marketplace is shaped by countless individual decisions by consumers and producers who are in continuous indirect communication with one another, consumers making their desires known via what they choose to buy and the prices they're willing to pay and producers making their costs known via the prices they charge.

Many students of markets would add a fourth sentence to this succinct description, something about preventing monopolies and ensuring energetic competition. We'll address competition in later chapters, but I omitted it from the above summary of basics because so many putative champions of markets praise competition in theory but undermine it in practice. Case in point: over the years, the members of the Federal Trade Commission, who could choose to assert that watchdog's potent antitrust powers, have only challenged between 1 and 3 percent of proposed mergers, with the percentage being lowest during Republican administrations. As a result, many massive corporations enjoy monopolistic advantages and exploit them vigorously. Ironically, while so many free market advocates have long ignored this violation of free market principle, for decades opposition to monopoly has come mostly from the political left.

Advocates for markets believe that supply and demand leads us close to the holy grail of economics: the optimal allocation of resources. In this Econ 101 scenario the purchasing decisions of consumers guide producers to make the best use of capital, labor, and raw materials to satisfy the wants of consumers while still ensuring that the producers make enough profit to stay in business. If consumers start buying more pizzas and fewer hamburgers, producers will start shifting more capital, labor, and raw materials from burger shacks into pizza joints. The producers will continue shifting resources into pizza ventures as long as consumer demand for pizza keeps growing. When the return on those pizza investments levels off, it means producers and consumers have hit the sweet spot, the balance point at which resources are being allocated optimally between pizza and hamburgers. Consumers win, producers win, and society gets the most value from its finite resources. Extrapolate from pizza and burgers to all goods and services and you have a market economy, a system organized by billions of unorganized individuals making billions of unorganized choices every day. The veneration accorded supply and demand is captured in an old economics joke (yes, they exist): "Catch a parrot and teach him to say 'supply and demand,' and you have an excellent economist."

Most sustainability economists appreciate the elegance of supply and demand. They further appreciate that in many situations it does indeed maneuver the economy close to the optimal allocation of resources, just as in the dreams induced by the reverent tones of the Milton Friedman Choir. But sustainability economists realize that in many other situations the supply-and-demand dream darkens, and all too often it leads to nightmares.

Frequently those nightmares result from the market's inability to detect social costs, one of our economy's most basic and pervasive flaws and the deficiency that the SCC seeks to correct in the realm of climate change (hence the term "*social* cost of carbon"). People not schooled in economics see the damage done by this defect in their everyday lives, but they have no name for it, no understanding of its origins, no sense of its magnitude, and no clear idea how to deal with it. This defect is no secret to economists, however. It makes at least a brief appearance in almost all introductory economics textbooks. But most neoclassical economists treat it as a minor matter whereas sustainability economists consider it a big deal.

The official term for this fundamental flaw is "externality." Economists recognize a technical difference between externalities and social cost, but in the context of the SCC and most environmental issues, these two terms are essentially synonymous, and in this book I will use them interchangeably.

At their core, externalities stem from a problem highlighted in the classic movie *Cool Hand Luke*. Near the end of the film, Luke escapes prison only to get trapped in an old church by his pursuers. Resigned to his fate, he moves to a window in full view of the guards and, echoing a remark directed at him earlier by one of the bosses, delivers one of Hollywood's most storied lines: "What we've got here is a failure to communicate." Immediately (spoiler alert) one of the pursuers shoots him, and he dies without uttering another word.

In the case of the economy, a failure to communicate doesn't produce quite such theatrical results, but the consequences most definitely include matters of life and death. And externalities often disrupt the communication between consumers and producers that enables supply and demand to work.

For a definition of externalities, let's turn to *Principles of Economics*, a widely used textbook written by Harvard professor N. Gregory Mankiw, a prominent conservative economist and the former chair of the president's Council of Economic Advisers under George W. Bush. Mankiw writes that an externality "arises when a person engages in an activity that influences the well-being of a bystander and yet neither pays nor receives any compensation for that effect."

Why does Mankiw use the word "influences" rather than "harms"? Because he is defining both negative and positive externalities. If a coal-fired power plant emits pollutants that harm people, that's a negative externality. However, if the smokestacks of that coal-fired power plant exhale hundred-dollar bills and if those Benjamins drift into the hands of nearby residents, that would be a positive externality. Exhalation of C-notes is wishful thinking, but actual cases of positive externalities exist, such as when a beekeeper raises bees for honey and they provide free pollination to a neighbor's orchard. However, when it comes to the environment, the vast majority of externalities are negative, so when I use the word "externality," I mean negative externality unless I specify otherwise.

Why does Mankiw use the word "bystander"? Because something is an externality only if it affects people who aren't part of a transaction. When a consumer buys electricity produced by a coal-fired power plant, the price the consumer pays for that electricity is not an externality; the consumer's payment is internal to the market arrangement between the consumer and the utility. However, if that consumer suffers from asthma attacks incited by pollutants from that power plant and has to buy asthma medication, the cost of that medication is an externality; her transaction with the power company doesn't include having asthma.

Why does Mankiw write that for an effect to be an externality the person or company causing that effect must neither pay nor receive any compensation for that effect? Because a bystander is no longer a bystander if she gets paid. If that person suffering from asthma made a deal with the power company in which she agreed to suffer from asthma in return for $100 a month, then the cost of her asthma meds is not an externality.

Note that producers of externalities may or may not be aware that they are generating social costs. Taking advantage of externalities is a conscious strategy for some enterprises, while for others it's unintentional. Note, too, that producers aren't the only people who benefit financially from externalities. Producers (and their investors) sometimes keep all or most of the extra profits attributable to uncompensated social costs, but other times competitive forces cause some of those ill-gotten gains to trickle down to consumers via lower prices. Producers bear primary responsibility for the ubiquity of externalities, but we consumers frequently are complicit, even if unwittingly.

THEY'RE EVERYWHERE

Now, in the name of advancing knowledge, I must make an embarrassing confession: I struggle with a particular thought process. Let's say I'm cutting the leafy tops off some fresh strawberries and dropping the shorn berries into a bowl of ice cream for a summer treat. Everyone I know who is old enough to be trusted with a sharp knife can do this while chatting, listening to a podcast, or watching *Jeopardy!* and naming the capital of Burkina Faso in the form of a question. I cannot, and not just because I don't know the capital of Burkina Faso. If I let my focus waver, I will sometimes dispose of the berry and put the leafy top into the ice cream. Seriously.

This problem is not limited to strawberries. If I crack open an egg, I need to be careful to drop the egg into the pan and get rid of the shell, not vice versa. If I strip the peel off a slice of orange, I will mix up which part to eat and which part to toss unless I concentrate. You see the pattern.

I don't know the source of my impairment, but I wonder if it doesn't stem at least partly from my childhood habit of sniffing gasoline. I loved the smell of gas. Whenever my parents filled up the old station wagon, I'd stand right next to the nozzle and suck those fumes up my nose as if savoring lilacs in bloom. The thing is, this was back when gasoline still contained lead. And inhalation of the toxic by-products associated with leaded gasoline was one of the main sources of lead poisoning before that fuel was banned.

When residents of Flint, Michigan, recently discovered that lead contaminated their drinking water, it vividly reminded Americans about the pernicious nature of this neurotoxin. Lead poisoning increases adults' risk for high blood pressure and kidney damage. Pregnant women exposed to lead are more likely to have miscarriages, stillbirths, and babies with low birth weights and minor malformations. But young children suffer most grievously.

It turned out that about 5 percent of Flint's children had elevated levels of lead in their blood. Subsequent studies revealed that Flint is not an outlier; a 2017 Reuters analysis found that 3,810 U.S. neighborhoods suffered from lead levels at least double those of Flint. According to a 2017 statement from the Centers for Disease Control and Prevention

(CDC), "Today at least 4 million households have children living in them that are being exposed to high levels of lead." Estimates of the number of kids who suffer from lead poisoning vary considerably, largely due to uncertainty as to which blood lead level to use as the threshold, but consider this: millions upon millions of American children have some amount of lead in their bodies and, as the CDC puts it, "no safe blood lead level in children has been identified."

Occasionally high levels of lead poisoning will put kids into a coma or kill them, but perhaps the most chilling widespread impact on kids is the damage that even low levels of lead do to their vulnerable developing brains. Lead can cause irreversible mental disabilities and behavioral disorders in kids. This reduction in their ability to think and control their behavior results in all sorts of misery for these kids and their families. It hits society in the pocketbook, too. Peter Muennig, an MD and professor in the school of public health at Columbia University, produced an eye-opening study on the costs of lead-related brain damage to children aged six and under. He focused on the increase in crime attributable to the behavioral problems caused by lead and on the loss of IQ due to lead poisoning. That IQ loss leads to lower educational attainment, and less education results in lower earning power, higher welfare costs, and higher health costs. Even with Muennig's narrow focus, his research indicates that if a child is poisoned by lead, the average cost to society over that person's lifetime will be about $50,000 extra.

The range of injuries caused by lead exposure suggests how varied and sometimes subtle social costs can be. Often the cause and effect are so removed from each other in time and geography that people don't connect the dots. Given their elusiveness and sheer number, it would be futile to try to compile a thorough list of examples. Instead, I'll tell you about a random twenty-minute walk I took through my hometown to provide a peek at the tip of the externality iceberg. Bear in mind that I live in Corvallis, Oregon, a city of some sixty thousand residents that enjoys a well-deserved reputation as one of the most sustainable communities in America, but even so . . .

As I began my stroll along a downtown sidewalk, I immediately spotted my first externality: the sidewalk itself. The main ingredient in the concrete beneath my shoes was cement, and the extremely energy-intensive cement industry produces about 5 percent of global greenhouse gas emissions—a staggering figure for a single industry.

Corvallis has more than its share of bicycles, electric and hybrid cars, and buses that run on alternative fuel. Still, fossil fuels powered the vast majority of the vehicles motoring past me, and the witches' brew blowing from those tailpipes causes all sorts of harm in addition to its impact on climate. Consider ground-level ozone. Most of us know it as an ingredient of smog, but it also slows plant growth, stunting crops and forests. Ozone even disrupts the navigational systems of bees, making it difficult for them to find flowers and carry out their vital function as pollinators. Of course, we mostly associate smog with health problems.

Social costs stemming from impacts to human health showed up literally around every corner during my walk, so rather than recite a litany of examples I'll just note a few broad findings, starting with a recent World Health Organization report, *Preventing Disease Through Healthy Environments*. This study estimates that about a quarter of the global disease burden and a similar fraction of all premature deaths stem from environmental factors, most of them externalities. More specific to this book's subject matter, health-related externalities tied to climate change hit the headlines in 2018 due to the publication of two major reports: the U.S. federal government's annual National Climate Assessment and a sweeping study published in *The Lancet*, one of the world's most respected medical journals. The spread of tropical diseases, huge spikes in heat-related illnesses, rising childhood malnutrition due to drought and flooding—the reports warned of mounting health risks in both the present and the future. No wonder that in 2019, seventy-four leading medical and health organizations singled out global warming as the century's "greatest public health challenge" and urged government to treat it as a "health emergency."

Continuing my stroll, I soon passed Robnett's Hardware on the corner of Adams and Second, where this business has stood since 1855. Many times have I trod on the store's massive floorboards, burnished by nearly a century of customers searching for hammers, screws, ladders, and wrenches. Back in the 1920s someone sawed the lumber for that floor from some of the hulking conifers that originally greened most of western Oregon. Much of the forest, especially the old growth, is gone now, cut for lumber or to clear land for development, and its absence creates numerous externalities.

Forests provide all sorts of services to us humans, but at the moment I'm thinking about 1996. I remember standing at the south end of downtown Corvallis in early February of that year, on the west bank of the Willamette River just a few blocks from Robnett's. As the heavy rain ricocheted off my jacket, I watched the chocolate-colored floodwaters rising, licking the toes of the commercial district. In the slightly lower neighborhood just south of downtown, residents kayaked along the drowned streets, passing inundated homes and stores. This turned out to be the biggest Oregon flood in decades, and it caused immense damage.

Many ingredients go into the pot to cook up major floods, but deforestation can play a significant role. For example, tree roots create spaces in the soil that can absorb water, but when forests have been cleared and replaced by parking lots or farm fields, runoff has nowhere to go but downslope to already brimming creeks and rivers.

Happily, on the early-spring day when I took my walk, the Willamette was flowing well below flood stage. But the river harbors other externalities besides heightened floods, including the damage to fish caused by logging, farming, and other industries. In pioneer times, healthy runs of salmon and steelhead periodically surged upriver, passing through Corvallis on the way to their spawning grounds. Locals could throw a hook into the water and reel in dinner—or dinners for a week if they caught a twenty- or thirty-pound Upper Willamette River Chinook. But development has decimated that wild abundance. Nowadays residents seldom snag a salmon from the river and are far more likely to go to the market and buy a Chinook fillet from Alaska, if they can afford $25 a pound. And those beefy Upper Willamette River Chinook? Well, now they're listed as "threatened" under the federal Endangered Species Act and are struggling to survive. A 2019 study found that they are especially vulnerable to climate change.

In an attempt to partly replace ailing wild populations, governments built salmon hatcheries throughout the Pacific Northwest. Hatcheries bring to mind another category of social cost, one easily overlooked: public efforts to mitigate some of the damage done by private enterprise. Mostly funded by taxpayer money, battalions of local, state, federal, and university workers dedicate huge amounts of time and resources to fisheries management and research. Such endeavors have existed longer than most people have been alive, so it's easy to miss the fact that this sprawling management network wasn't needed prior to dam building, overfishing, excessive pollution, and habitat degradation. This category of social cost reaches well beyond fisheries and includes the management of logging on public forests, grazing on public lands, and recreation in public parklands.

The plight of those imperiled Upper Willamette River Chinook represents one of the world's most important externalities: the loss of biological diversity. The term often evokes pandas, koalas, and other cute animals we like to watch on YouTube, but biodiversity plays a consequential role in human well-being. The list of benefits provided by biodiversity is long: a genetic storehouse that enables the development of new crops, the building blocks of ecosystems and the essential services they provide, and a source of new medicines. Similarly long is the list of human activities that degrade biodiversity. In 2018, the executive secretary of the UN Convention on Biological Diversity warned that the loss of biodiversity is as perilous as climate change and that humans could go extinct if we continue to snuff out species at the current rate. In 2019, a draft summary of a UN report on biodiversity made headlines around the world. For an extreme summary of the summary, here's part of what Robert Watson, chairman of the body that produced the report, told *The Guardian*: "There is no question we are losing biodiversity at a truly unsustainable rate that will affect human wellbeing both for current and future generations. We are in trouble if we don't act." The report estimates that we're losing species at a rate one hundred to one thousand times faster than the natural rate, and that around one million species will forever vanish within the next few decades unless we humans make profound changes. The researchers note that climate change is a major and increasing cause of biodiversity loss, but other factors, such as habitat loss and invasive species, also drive these mass extinctions.

Soon my route through downtown took me past the Shell gas station. Air pollution, oil spills, climate change—even though gas stations no longer dispense leaded gasoline that some clueless kid might stand around and sniff, they overflow with externalities. Let's consider the less well-known social costs of the oil supply chain, such as the many harmful effects resulting from the loss of hundreds of square miles of wetlands along the Louisiana coast, due in large part to oil and gas development. Most dramatically, the now-tattered shield of wetlands provides less protection from storm surges than in decades past, contributing to disasters like the flooding of New Orleans during Hurricane Katrina. Ironically, the rending of the coast's natural shield also threatens some $100 billion worth of oil and gas infrastructure. Quoted in a 2016 Bloomberg article, Kai Midboe, director of policy research and general counsel at the Water Institute of the Gulf, said, "The industry down there has relied on the natural environment to protect its infrastructure, and that environment is now unraveling."

I made the final stop of my whirlwind tour at the Safeway. As I walked the aisles, externalities pressed in on me from every side, as they would in almost any store of any sort once you take supply chains into account. Consider the strawberries. Most nonorganic growers spray strawberries with lavish amounts of pesticides. A 2019 study in *Biological Conservation* found that global insect populations are in sharp decline, mostly due to habitat loss from industrial agriculture and the widespread application of pesticides (with a significant assist from climate change). We may joke about not missing mosquitoes or cockroaches, but the fact is people cannot survive without insects.

Given the abundance of externalities in the store, trying to enumerate them was futile, so instead I headed for the seafood section in search of an exceptionally heinous social cost. Within a minute I'd found my quarry: shrimp from Thailand.

I can't say for sure that slaves produced the particular shrimp I spotted, but much of the shrimp Americans eat comes from Thailand, and some of the shrimp from Thailand is peeled and gutted by workers who are literal slaves. A Pulitzer Prize–winning investigation by the Associated Press, as well as numerous government and public-interest organization reports, found that many workers in the Thai seafood industry are rounded up by human traffickers and forced to labor behind locked doors, sometimes for sixteen hours a day and seven days a week for years. Escaped workers report severe beatings, putrid living quarters, and brutal working conditions that often leave workers sick and injured. Among these slaves are many children; AP investigators saw one girl so young that she had to stand on a stool to reach the peeling table.

The AP investigation revealed that massive amounts of shrimp processed by slaves entered the supply chain that eventually put cheap shrimp on the shelves of most major American supermarkets and restaurants, including behemoths such as Walmart, Kroger, Red Lobster, Olive Garden, and, yes, Safeway. In fact, a photo of alleged illicit shrimp used in one AP story showed a package of Waterfront Bistro shrimp just like the package I found in the Corvallis Safeway. In the wake of the AP revelations, Safeway and many other retailers vowed to ramp up their efforts to remove the slavery links from their supply chains, so that package I held may have been slavery-free. Thailand also has recently made significant improvements, according to David Pinsky, the senior oceans campaigner for Greenpeace USA. However, says Pinsky, government and media reports indicate that slavery shrimp still ends up on our dinner plates all too often; the appetite for cheap shrimp makes it difficult to root these abuses from the multibillion-dollar seafood market.

Pinsky produces Greenpeace's annual report ranking supermarkets according to the sustainability of their seafood, which includes a section on slavery. In our conversation he emphasized that Thai shrimp is just one example of a much larger problem that encompasses seafood in general and includes dozens of countries. Slavery on the high seas seems to be worst of all. "If you can have these instances of child and forced labor happening in a peeling shed on land, then imagine what's happening thousands of miles out there away from shore," said Pinsky. Investigations of these slave ships have produced gut-wrenching reports of torture, murder, and people kept in cages.

A high price to pay so producers and consumers can enjoy low-priced seafood. "Externality" is such a bloodless term for such a bloody market failure.

MARKET CORRECTION

Okay, so we're awash in externalities. What does that have to do with the basic structure of our economy?

Well, consider gasoline. Let's say a guy drives into a gas station and fills up with regular for about $2.50 a gallon, the average price at the time I wrote this paragraph. That's a good price; adjusted for inflation, it's cheaper than gas usually has been in America. With gas so cheap, this guy recently upgraded to a roomy Cadillac ATS sedan. He even went for the model with the 3.6-liter V6 engine, despite the fact that it gets only eighteen miles to the gallon around town. But what the heck; mileage hardly matters when gas costs several times less per gallon than bottled water, right?

Perhaps before buying that Caddy our guy should have looked at some of the research conducted by Milton Copulos. Copulos was a resource security expert who served as a consultant to the Reagan administration, as an adviser to Reagan's CIA director, and as the director of energy studies for the Heritage Foundation, a heavyweight conservative think tank. Concerned about the hidden costs of America's dependence on imported oil, Copulos gave testimony on this subject to the Senate Foreign Relations Committee in 2006. Early on he said, "The principal reason why we are not fully aware of the true economic cost of our import dependence is that it largely takes the form of what economists call 'externalities' . . . It is important to understand that even though external costs or benefits may not be reflected in the price of an item, they nonetheless are real." He went on to detail the hidden costs he and his associates had calculated, which consisted mainly of military expenditures and oil supply disruptions. Then he hit the committee with the punchline: if these externalities were included, the true cost of imported gas would rise by $5.04 a gallon. Using our example of $2.50 a gallon, that means the price should be $7.54 a gallon.

Many other researchers have examined many other externalities associated with gasoline, such as the health impacts of smog, reduced crop yields due to air pollution, and higher insurance rates due to increased extreme weather events resulting from climate change. Their results vary, but like Copulos, they tend to put the true cost of gasoline somewhere between double and quadruple the current price at the pump. Most of the researchers also acknowledge that many externalities are tough to calculate and were omitted from the research, so their estimates are almost surely low.

What would happen if prices accurately captured the real costs of production and consumption, which, according to standard market theory, is critical to the efficient functioning of the market? Well, if gas cost $7.54 a gallon, our guy who bought the new car probably would not have opted for the 3.6-liter V6. He probably would not even have bought a Cadillac ATS. More likely he would have chosen something much less expensive so he could afford $7.54-a-gallon gas. Of course, if gas had been fully and accurately priced since the days of the Model T, by now our guy probably would have been able to choose among hundreds of all-electrics and who knows what other kinds of non-gasoline-powered vehicles. Or maybe he would have been living in a compact, walkable community with an excellent transit system and he simply would have bought a new pair of walking shoes and a rail pass.

"People are very rarely aware of the sort of externalities that come up," said Dean Baker, the cofounder and senior economist at the progressive Center for Economic and Policy Research. "There are probably a lot of cases in which people would make different choices if they were aware of [the externalities]." Unaware of the social costs, notes Baker, consumers often just choose the cheapest product. "But if they found out that the lowest cost item was made under very bad environmental conditions or really exploited labor or was an anti-union place or used child labor, a lot of people would choose not to buy it. But in general you don't have that information and it would take you a while to get it. So you end up making a choice you wouldn't make if you were better informed."

Even though most mainstream economists don't seem to recognize the degree to which social costs foul the machinery of the market, almost all economists at least recognize the reality of social costs. Almost all. A few free marketeers take a more extreme stance. For instance, the Cato Institute, an ardently laissez-faire think tank, published a slim book by economist Steven N. S. Cheung entitled *The Myth of Social Cost*. Underscoring the doubts initiated by the word "myth" in the title, the book often uses quotation marks to further sow skepticism. For example, the first sentence of the prologue reads, "Society might be far better off if the 'problem' of social cost had never been discovered," implying that social cost is not a real problem. Equally loaded quotation marks pop up in the prologue's last sentence: "The supposed existence of 'social cost' has been one of the foremost pretexts for which such freedom has been transgressed and by which the authority of government has been extended." This last sentence reveals the main concern felt by many people who deny or downplay social costs: if social costs exist, government might try to reduce those costs, and the deniers deeply dislike government intervention in the marketplace.

John Charles, president and CEO of the Cascade Policy Institute, a free market think tank, echoed Cheung's antigovernment theme when I spoke to him, yet he readily acknowledged social costs. Apart from "hello," pretty much the first words he said to me were "Externalities do exist." He went on to note numerous examples, revealing a particular hostility toward the noise produced by leaf blowers. (He was joking—mostly.) His main concerns centered on what he sees as the folly of government trying to deal with externalities via broad mechanisms like the SCC. He sees SCC-like approaches as ineffective and largely designed to raise taxes. "You just stick on another tax and the money goes into some pot of money. I've been a legislative lobbyist long enough to know that once you put money on the table, you can forget it, it's game over. Someone's going to get their straw into that pool of money and it's going to be diverted into some stupid thing."

Despite his dim view of government, Charles is open to regulation in some circumstances. As a way to address a specific, limited social cost, he prefers a specific, limited regulation, such as stopping a factory from dumping toxic waste

into a river, to an economy-wide intervention such as the
SCC. However, when it comes to climate change, Charles
says we should "just do nothing" because there's no real problem. "In the case of CO_2 there's never been any definitive
evidence showing that there's a causal relationship between
human activity and changes in global climate. What is global
climate? What are changes? Average temperature? What is
that? Who cares?"

Let's return to Charles's previous point about the danger
that complex, systemic approaches to social costs will lead
to more taxes. This could well be true, but many people see
taxes as a necessary solution and don't feel as cynical about
how that public revenue would be used. Basic economics does
recommend taxes as one of several remedies for externalities.
Such levies are called "Pigouvian taxes" (also spelled "Pigovian") after Arthur Pigou, an early and mid-twentieth-century
British economist who expanded the concept of externalities
and brought it into modern economics. Pigou figured that
government could assess a tax roughly equal to a social cost
and thereby internalize the externality, creating accurate market signals, à la our earlier discussion of establishing a true
price for gasoline.

Many neoclassical economists view such internalizing as a
pro-market approach. In fact, some conservatives hail internalizing social costs as the best way to correct market failures and thereby stave off regulation, enabling the economy
to remain mostly laissez-faire. Most sustainability economists
agree only to a point. For one thing, they recognize the abundance of externalities and question the practicality of trying
to internalize them all. Still, many sustainability economists
do think it would be immensely helpful to internalize some
important externalities as much as possible. For a prime
example of an externality that many people would like to

internalize, you need look no further than the social cost of carbon. When used as the basis for pricing carbon, the SCC represents a classic effort to internalize an externality.

N. Gregory Mankiw, the Harvard professor and former adviser to George W. Bush whose textbook definition earlier introduced us to externalities, numbers among the conservative economists who see internalizing social costs as a pro-market solution. Mankiw even founded the Pigou Club, which he describes as "an elite group of economists and pundits with the good sense to have publicly advocated higher Pigovian taxes, such as gasoline taxes or carbon taxes." The membership mostly consists of economists spanning the philosophical spectrum, though their agreement about Pigouvian taxes goes only so far; they differ on important details, such as how high such taxes should be or how the revenue should be spent. In addition to the econs, a varied array of other famous people also have made the list, including Al Gore, Rex Tillerson, Ralph Nader, Lindsey Graham, and Bill Gates. True oddities pop up here and there, too, such as actor Jack Black.

The Pigou Club is not an actual club, I'm sorry to say. There is no wood-paneled room in Manhattan where Tillerson and Black sip brandy together. Mankiw created this fictional assemblage and populated it with people who at one time or other have made known their support for Pigouvian taxes, even if the non-economists among them may not even know the term. Mankiw was simply demonstrating the wide range of advocates for such taxes. A rival gang, the NoPigou Club, came together in opposition, but it fizzled before a Jets-and-Sharks-style street brawl could erupt between rival economists.

Mankiw largely attributes opposition to Pigouvian taxes to the ignorance of the general public, whom he playfully refers to as "Muggles" in at least one paper. He chastises politicians, as well, but gently because he figures their reluctance to champion Pigouvian taxes generally stems not from ignorance but from what he sees as their justifiable fear of being pitchforked by tax-phobic mobs. I think he lets the politicians off too easy, particularly his conservative brethren, who have been demonizing taxes for decades and bear much of the responsibility for the knee-jerk anti-tax reflex that sends so many citizens looking for their pitchforks.

A few years ago I spent some time in Washington, D.C., talking to economists, think tank denizens, and federal agency researchers. After a week of geek, I decided to seek some perspectives distant in geography and culture, so I shifted gears and headed a couple of hundred miles west to coal country in West Virginia. Because coal ranks as a huge source of carbon dioxide emissions, the already declining coal industry could take a sizable hit if the government uses even a modest social cost of carbon in its policy-making, which seemed likely when I made this journey in pre-Trump times.

At the time of my West Virginia trip many locals were worked up about the so-called war on coal. I could have talked to any number of irate West Virginians who blamed their plight on what they perceived as attacks by former president Obama and his administration's environmental regulators. If I had talked with them about the federal social cost of carbon, which would add to King Coal's economic woes, I expect some of them would have stormed off immediately and others might have stuck around just to give me a good tongue-lashing. It's hard to calmly consider the social costs your bread-and-butter industry is imposing on others when you and your own community are suffering.

Fortunately, I got to talk with Jim Petitto. Make no mistake: coal runs through Petitto's veins. He's the president of a small family-owned business in Morgantown that invents, manufactures, and sells a specialized line of coal-mining equipment. Petitto's dad started the company in the 1960s, and a third generation of Petittos is now helping run the family business. "We really grew up in a coal culture," he said, and he shares much of that culture's antipathy toward Obama and the pre-Trump federal government. But Petitto recognized the complexities of the situation and pondered the issues in an open-minded and thoughtful way.

I met Petitto at the modest office of Petitto Mine Equipment. Scattered around the adjacent gravel yards were some Petitto Mules, the company's main product line. These husky, low-slung vehicles run on treads like tanks and can weigh up to fifty tons. Petitto wore a cap bearing an image of a Petitto Mule, topping off his ensemble of worn jeans and a T-shirt reading "Life Is Good." His casual outfit matched his friendly demeanor, and he spoke with a slight West Virginia drawl that further reinforced his easygoing manner.

Petitto began by telling me about his family and company history. He talked about how his dad grew up in a coal company town and worked as a coal miner before he started the business and invented the first Petitto Mule; about all the long hours and hard work his family put in; about the fact that a fair number of their several dozen employees had been with Petitto Mine Equipment for thirty and forty years. He also spoke about how gratified he was that Petitto Mules had made underground coal mining faster and, more important, safer. "We're a really small company, but we're very proud of what we've accomplished."

Petitto said that his company was doing well, but many of his fellow West Virginians were not. "It's just heartbreaking, absolutely heartbreaking," he said, referring to the small towns that have been dependent on coal for generations. "You go by the [shut-down] mine sites, and they have the helmets, the hard hats, on crosses." He pointed out that many of these little towns are remote, so ex-miners can't just commute somewhere for work, even if there were jobs. "There's no place to go, nothing to do."

When I brought up the social cost of carbon and externalities, Petitto readily acknowledged the reality of the concept. He also said that "climate change is definitely occurring" and that greenhouse gases contribute to the problem, though he added, "But I do believe in some of the arguments that there's a [natural] cycle going on, too. I think it's a combination of things."

Faced with the battling realities of the harm coal causes and the benefits it brings to his neck of the woods, Petitto felt the pain of the dilemma. Several times he mentioned wanting to "do the right thing" for the environment, once saying, "I don't want to be part of an industry that says blatantly, 'Ah, the hell with it, we gotta burn coal.'" But at least as many times he expressed concern over the economic and social problems that people in coal country would face if the coal industry continues fading. He largely pinned his hopes on clean coal technology, thinking it could enable us to continue burning coal without adding much to climate change or other air pollution problems. The idea of applying the SCC and establishing a carbon tax worried him—"Just a flat-out carbon tax would probably cripple the coal industry, if it could be crippled even more"—but he thought it might work if some of the revenue from the tax was plowed into advancing clean coal.

Sadly, clean coal technology appears to be a mirage, always beckoning but forever receding out of reach. Despite repeated attempts, no one has yet made the technology commercially viable. Nor does it seem possible that President Trump's promise to "bring the coal industry back one hundred percent" can be kept, not with the market forces and environmental realities confronting the coal industry. But talking with Jim

Petitto and hearing his distress over the struggles in coal country serve as a moving reminder that the changes required to diminish the social costs of carbon, to reduce emissions and switch to a clean energy economy, come with a human price. The fact that making the necessary changes to address externalities will help far more people than it hurts doesn't relieve us of the responsibility to help those faced with a rough transition.

2

THE MOST IMPORTANT NUMBER

YOU'VE NEVER HEARD OF

In 2010, the SCC surfaced as a whisper, deep in the federal bureaucracy. It debuted in the "Final Rule Technical Support Document (TSD): Energy Efficiency Program for Commercial and Industrial Equipment: Small Electric Motors," Appendix 15-A. Squint at the table on page 10,910 of the March 9, 2010, *Federal Register*, and you'll spot the number "21" in a font size that will challenge your eyesight.

Twenty-one was the original U.S. federal government estimate for the social cost of carbon. This means that a ton of carbon emitted in 2010 was expected to cause $21 in damages during its time in the atmosphere. Informing federal regulations was the primary motivation behind the development of the SCC, driven by President Ronald Reagan's executive order that requires regulators to assess the costs and benefits of proposed rules to help guide decision-makers.

Despite its potential as a game-changer, for years few people knew about the SCC, prompting some insiders to refer to it as "the most important number you've never heard of." It still seems unlikely that the SCC will push "Bigfoot Kept

Lumberjack as Love Slave" and its ilk off the front pages of the tabloids, but its profile has risen a bit. The SCC has shown up in precedent-setting court cases, played a part in some high-stakes political tussles like the Clean Power Plan, and drawn fire from fossil fuel interests and conservative think tanks. That heightened attention from industry and the political right likely explains why President Trump axed the working group that developed the SCC shortly after assuming office and why his administration has tried to kill or cripple the SCC itself.

By calculating the social costs of a ton of carbon dioxide, the SCC is ultimately estimating the benefits associated with *not* emitting that ton of CO_2. The costs in cost-benefit analysis refer to the costs to industry. For example, in the regulation mentioned above, industry would shoulder the cost (if any) of producing more energy-efficient small electric motors. It would make no sense to impose this cost unless society gained benefits that outweighed those costs. But determining costs and benefits is knotty and often skewed by biases. Some of those biases are simple, self-serving attempts to game the process, such as when industry tries to pump up the estimated costs of a proposed rule by exaggerating the expense of compliance.

The even more distressing biases are structural, notably the fact that the costs to industry typically are more obvious and more easily measured than the benefits to society. The social rewards of a stable climate provide a prime example of a complex and elusive benefit. The basic purpose of the SCC is to assign a dollar amount to the benefits of a stable climate so that an apples-to-apples comparison can be made when weighing those benefits against the costs of reducing greenhouse gas emissions. That's why the fossil fuel industry generally opposes the use of an SCC or tries to keep the number as low as possible, because the smaller the projected benefits, the less likely a regulation will be adopted. As we'll see in the course of this book, navigating the costs and benefits of global warming is a fraught venture with repercussions that reach far beyond regulations.

One such repercussion is the possibility that the SCC will play an important role in determining a price for carbon—another reason the fossil fuel industry wants a diminutive SCC. This worries many advocates of strong climate action because they think the price derived from a flawed or even rigged SCC process would be too low to significantly change consumer behavior. For example, by some estimates a carbon price of $21 would raise the cost of gas by only about 20 cents a gallon, hardly enough of an increase to motivate many people to drive a lot less or to buy an electric car. (Periodic adjustments during the Obama administration gradually increased the SCC to about $40, but the Trump administration dropped the number to $1 to $6, a cripplingly low range that faces legal challenges.)

The group of experts who produced the original SCC understood that it was a rough approximation. In fact, due to all the uncertainties, the group developed a range of numbers, each depending on a different set of assumptions and inputs. Twenty-one was the central estimate, the figure that typically gets used in regulations, policy, and political debate. Imprecision notwithstanding, the government touted the number and the process as a solid beginning for the SCC, a view shared by many observers and later legitimized by several court decisions.

Still, the SCC has plenty of critics. Predictably, fossil fuel interests have lambasted the SCC as—and I'm paraphrasing here—an economy-killing, tree-hugger-pandering, freedom-hating, fact-free farce. Political centrists and leftists have also questioned the SCC. Some think it can play a leading role in the fight against climate change if it is improved and used judiciously. Others think it has value but should play only a modest role. A few think it should be kicked to the curb, as evidenced by remarks from two well-regarded climate-friendly economists, one of whom called the SCC "an exercise in absurdity" and one of whom labeled it a "wild-ass guess."

How did this provocative number come to be?

In 2007, when the judges in the *Center for Biological Diversity v. National Highway Traffic Safety Administration* case ruled that the social cost of carbon can't be zero, events and legal decisions already had forced the George W. Bush administration to take a few baby steps toward reckoning with climate change. Notorious for wreaking environmental havoc, the Bush administration reluctantly initiated a number of modest climate actions, including one directing some executive agency staffers to develop an SCC. They made a little progress, but the halfhearted effort never went far.

Early in 2009, shortly after President Obama took office, his administration launched a much more energetic effort to tackle global warming and the SCC rode along on the tide. Michael Greenstone, chief economist of the president's Council of Economic Advisers, and Cass Sunstein, a prominent legal scholar who headed the obscure but powerful Office of Information and Regulatory Affairs (OIRA), reportedly sat down in the White House cafeteria one day and hatched the idea of establishing a unified SCC. Prior to 2009, various federal agencies had created and used some widely varying numbers, but Greenstone and Sunstein felt the government needed a consistent, scientifically defensible SCC.

By the spring of 2009, Greenstone and Sunstein had formed the Interagency Working Group on Social Cost of Carbon (IWG) and directed it to hammer out the SCC. (Whenever I use the acronym "IWG," I'm referring to this entity, not to any of the government's many other interagency working groups.) The IWG consisted of representatives, most of them with an economics background, from a wide variety of federal offices and agencies. Picture a couple of dozen people sitting around a table in a nondescript Washington, D.C., conference room confronted with the task of taking our vast, complicated, and evolving knowledge of climate change and blending it with our vast, complicated, and evolving knowledge of the economy to determine how much climate change would cost us. And wrap it up by Christmas.

Clearly, the IWG members didn't have the time and means to trek across Alaska measuring glacial melt or to conduct surveys in Florida to find out how much sea level rise was costing coastal homeowners. The people in that conference room had to rely on existing scientific and economic research. But even gathering, sorting, synthesizing, and extrapolating from the existing research would have been overwhelming. Out of necessity, the group turned to IAMs.

No, not the pet food—that's "Iams." When it's all caps, "IAM" is the acronym for "integrated assessment model." Basically, IAMs blend knowledge from more than one discipline into a single framework. In the case of climate IAMs, the models establish various scenarios involving factors such as anticipated global population and technological changes. Then they estimate how much carbon dioxide would be emitted in each scenario, predict how much those emissions would raise global temperatures, determine how those raised temperatures would affect the planet and the people on it, and, finally, quantify those effects, usually by calculating the impact on gross domestic product (GDP). People who use the IAMs, such as the IWG, can go with the results found by the creators of the models (the "default" findings), or they can plug in some of their own scenarios and assumptions and then run the models. But even when users make some adjustments, the IAMs exert a huge influence on the SCC. Inevitably, these models have stirred passionate debate.

THE BIG THREE

Many climate IAMs exist, but I'll mention the Big Three: the Dynamic Integrated Climate and Economy model (DICE), by William Nordhaus; the Policy Analysis of the Greenhouse Effect model (PAGE), by Chris Hope; and the Climate Framework for Uncertainty, Negotiation, and Distribution model (FUND), originally by Richard Tol and in its later years a collaboration between Tol and David Anthoff. I've dubbed them the "Big Three" because they're the best known climate IAMs, they've been around for a long time, and they're the trio of IAMs enlisted by the IWG.

Nordhaus is a seminal figure in the world of climate IAMs. On the faculty of Yale University since 1967 and laden with honors, including the 2018 Nobel Prize in Economics, Nordhaus started working on the economics of global warming in the 1970s, long before most people had even heard of global warming. In the early 1990s he developed DICE, which he has revised several times over the years. The version of DICE used by the IWG produced a range of estimates suggesting that climate change posed a significant but not urgent threat. His numbers indicated that the costs of strong and immediate action would outweigh the benefits. Instead, DICE supported a course of moderate action at first with a gradual escalation of emission reductions over the decades. In recent years Nordhaus has done some recalculating and has called for more vigorous and rapid reductions, but his estimates still seem too tepid given the urgency of climate change.

Like Nordhaus, Chris Hope is a venerable figure in the field of climate economics; he began working on PAGE in the early 1990s. Also like Nordhaus, he sports an imposing résumé. Until his retirement, in 2018, Hope was on the faculty of the esteemed Judge Business School at the University of Cambridge. Hope and PAGE played a major role in the groundbreaking Stern Review: a seven-hundred-page, blockbuster report published in 2006 by the British government that provided an in-depth look at the economic impacts of climate change. The version of PAGE used by Stern and the IWG produced a wide range of SCC estimates, but the IWG overlooked the higher estimates and focused on the lower ones, settling on a central value similar to that of DICE. Later Hope collaborated with other climate economists and came up with SCC estimates several times higher than the figure employed by the IWG. Hope has long been recommending forceful and immediate action to curb greenhouse gas emissions.

Now, where to begin with Richard Tol? Perhaps we should start on the surface. Imagine that people are shown headshots of Nordhaus, Hope, and Tol from 2009 and asked to guess which two are eminent scholars and which one is a castaway just returned from years alone on a desert island. With his bushy beard and disheveled hair, the 2009 Tol would get 100 percent of the castaway votes. (When I recently spoke to Tol, his appearance had evolved to semi-castaway, but he still had the look of a free spirit.) Sure, appearances often prove misleading, but in this case they accurately convey the fact that he is an iconoclast. This shows up most significantly in FUND's bottom line; its default position indicates an SCC far lower than those calculated by DICE and PAGE. The following passage from FUND's website also conveys a bit of Tol's attitude: "An integrated assessment model, FUND is used to advise policy makers about proper and not-so-proper strategies. The model, however, always reflects its developer's world views. It is therefore regularly contrary to the rhetoric of politicians, and occasionally politically incorrect."

Born in the Netherlands, Tol holds professorships at the

Vrije Universiteit Amsterdam and the University of Sussex, in the United Kingdom. Over the years he has played a leading role with the Intergovernmental Panel on Climate Change (IPCC), though in 2013 he withdrew from a group responsible for writing a major IPCC report, accusing the international body of exaggerating the problem of global warming to serve a political agenda. Tol hasn't had as long and distinguished a career as Nordhaus or Hope—he is, after all, much younger—but he is a very influential climate economist. However, both his tone and the substance of his work have roiled the realm of climate economics.

A scan of Tol's blog, which he calls "Occasional thoughts on all sorts," quickly reveals his tone. Take his June 18, 2015, post, which criticizes Pope Francis's renowned encyclical on climate change and inequality. Tol allows that the encyclical contains both good and bad things, but his lengthy post dwells on what he considers the bad. Tol asserts that one paragraph of the encyclical "offers the alarmist claptrap you would expect to find in a Greenpeace magazine," that other paragraphs are "the stuff of failed undergrad essays," and that "the passages on intergenerational justice are mostly waffle."

I've gone into detail about Tol's sometimes combative style because it seems connected to his sometimes inordinate views of climate economics. Given FUND's frequent use in the development of climate policies, his views matter a great deal. The IWG accorded FUND the same weight as DICE and PAGE, despite some key findings in FUND that drifted outside the scientific-economic consensus. For example, the science community has urged keeping the temperature rise to less than 1.5°C above the preindustrial level and warned of dire consequences if the increase exceeds 2°C. Yet early versions of FUND, including the one used by the IWG, estimated that for decades the world likely would experience net benefits from warming of up to 2°C. An expert close to the

IWG process said of Tol's result: "I would call it lunatic. More polite people would call it counterintuitive."

Tol is not a skeptic about the scientific reality of global warming (though he thinks some people exaggerate its severity), but he holds an out-of-the-mainstream view of its economic impacts, a view some climate skeptics have commandeered. For instance, they have feasted upon Tol's projections that global warming will probably produce net economic benefits for many decades and only relatively mild costs in the long run. Providing such ammunition for their battles against climate action has brought Tol into the skeptics' orbit at times. Invited by coal giant Peabody Energy, Tol appeared as a witness to discredit the use of the federal SCC in a court case in Minnesota, a case Peabody unanimously lost. In 2015, Tol appeared in a climate-denying documentary called *Climate Hustle*, produced by the infamous climate skeptic Marc Morano and funded by the Committee for a Constructive Tomorrow, a group in turn funded by corporations like ExxonMobil and Chevron as well as conservative foundations and dark money donors.

When I recently asked Tol about his popularity among skeptics, he said that both sides of the climate debate appropriate his work, cherry-picking whichever results support their arguments. He noted that FUND, like all the IAMs, produces a range of estimates, and he said that the opposing sides choose the low end or the high end, whichever suits them. He pointed out that the assumptions and data that go into the model are flexible, interchangeable. "It's a framework for us to think through the problem. I don't see it as a rigid thing at all. It's a tool that you can use many different ways. You can swap assumptions, you can swap modules, you can swap parameters, you can swap data sets and see what happens." Tol is right about the ability of the model's users to make adjustments, but that overlooks FUND's default results, which are determined by his parameters and assumptions, and which generate a much lower SCC than other leading models.

Recent versions of FUND have become more moderate, perhaps due to the influence of David Anthoff, the model's coauthor in recent years. "He has a different perspective on things than I have," said Tol. "David has essentially reprogrammed the model and added some conceptual things, too." With a laugh, he added, "I'm not even sure that FUND reflects my thinking [anymore]."

To better grasp the connection between IAMs and the real world, let's return to the San Joaquin Valley during its recent drought. Envision those dusty, fallow fields I described in the introduction, fields that during wet years produce a wealth of fruits and vegetables. How does a climate change impact like those barren fields find its way into an IAM?

"Agriculture is a good example of what we [modelers] do," said Tol. "There are thousands of studies that estimate the effects of climate change on particular crops in particular locations under particular scenarios." These are the studies that involve detailed research, maybe someone plowing through precipitation records for a county in the San Joaquin Valley and correlating them with fluctuations in that area's tomato production. Such studies form the headwaters for IAMs. From there the tributaries of information flow into ever-greater streams, eventually feeding into state-, nation-, or world-sized rivers of data.

Calculating the overall impact of climate change on agriculture sets the table for the creators of the climate IAMs. "It is only at this point that we come in," said Tol. Presented with information from all those researchers, the IAM developers aggregate the results to get an overall estimate for agricultural impacts due to climate change. Then the modelers gather similar aggregated results that estimate global warming's effects on coastal real estate, fishing, transportation, human health, and myriad other sectors. Finally, the modelers weave all these threads together to form the substance of a climate IAM, which sets the stage for the number crunching that pops out an SCC number (or a range of numbers).

IAM and SCC developers readily acknowledge that their processes and results fall well short of perfection. But before we dig into those imperfections, we should take a moment to appreciate the achievement the SCC represents. It puts a value, however modest or inexact, on some of the externalities associated with climate change. This is a concrete effort to start correcting the greatest market failure of our time. The SCC also constitutes an acknowledgment that the market alone can't deal with global warming and other environmental and social goods.

LET'S GET BIOPHYSICAL

In 2010, when the IWG unveiled its original SCC, FUND's default central estimate suggested that the social cost of emitting one ton of carbon dioxide was $6.

But wasn't the first SCC $21? Correct, but that calculation included the central estimates from all the Big Three climate IAMs as adjusted by the interagency working group. After being customized by the IWG, PAGE produced a central estimate of $30 and DICE $28. So where did $21 come from? Simple. Add 6 and 28 and 30 and you get 64; divide by three and round off and you get 21, the average of the three numbers. The glaring gap between FUND's single-digit estimate and the results of the other two IAMs leads us right to a key distinction between neoclassical economics and sustainability economics: in the former, biophysical reality gets short shrift; in the latter, biophysical reality rules.

In 2009 and 2010, as word got around about the IWG's

work on the social cost of carbon, scientists expressed concern that the IAMs didn't adequately incorporate many aspects of biophysical reality. Some thought the IWG had gotten off to a good start but needed to improve its scientific underpinnings. Others thought the IWG had made a total hash of the science. The outcry led to a couple of SCC workshops—one in late 2010 and one in early 2011—that included a lot of scientists. Many of them delivered the same basic message: climate IAMs needed to better reflect the real world.

Consider the sizable weight FUND gave to carbon dioxide fertilization, which accounts in large part for the model's low SCC estimate. An infusion of CO_2 can ramp up photosynthesis and lead to increased plant growth, including the plants we humans consider food—a positive externality seized upon by opponents of climate action to bolster their assertion that climate change will produce net benefits to society (assuming, some add disdainfully, that climate change is even real).

Surely the most assiduous advocate of the notion that more carbon dioxide means more benefits is Roger Bezdek, a longtime consultant for energy companies. In 2014, under the auspices of an industry group called the American Coalition for Clean Coal Electricity, he wrote his magnum opus: *The Social Costs of Carbon? No, the Social Benefits of Carbon.* This 182-page paper found that the benefits of pouring carbon dioxide into the atmosphere far, far, far outweigh the social costs. According to Bezdek, the benefits are about fifty to five hundred times greater than the costs.

Most of Bezdek's benefits fall into a category he calls "indirect." Essentially he makes a heroic leap of logic, asserting that emitting carbon dioxide is a necessary component of economic growth given that fossil fuels drove the growth of the last couple of centuries. Bezdek's broad claim about the indirect glories of carbon dioxide is flagrantly bogus and doesn't merit detailed attention. More plausible and more salient to our discussion is his calculation of the "direct" benefits of emitting carbon dioxide, which emphasizes the flourishing of crops due to CO_2 fertilization. In his paper he sings the praises of carbon dioxide: "It is the primary raw material or 'food' utilized by the vast majority of plants to produce the organic matter out of which they construct their tissues, which subsequently become the ultimate source of food for nearly all animals and humans. Consequently, the more CO_2 there is in the air, the better plants grow . . . and the better plants grow, the more food there is available to sustain the entire biosphere." Bezdek estimates that CO_2 will deliver a $10 trillion boost to forty-five crops between 2012 and 2050.

Bezdek largely gets these numbers from "The Positive Externalities of Carbon Dioxide," a paper written by Craig Idso, founder and chairman of the Center for the Study of Carbon Dioxide and Global Change and a prominent climate skeptic. Here's a quote from the center's "about" section summing up its position on global warming: "There is little doubt the carbon dioxide concentration of the atmosphere has risen significantly over the past 100 to 150 years from humanity's use of fossil fuels and that the Earth has warmed slightly over the same period; but there is no compelling reason to believe that the rise in temperature was caused primarily by the rise in carbon dioxide. Moreover, real world data provide no compelling evidence to suggest that the ongoing rise in the carbon dioxide concentration of the atmosphere will lead to significant global warming or changes in Earth's climate." Nearly every climate scientist on earth would disagree with the center's position.

Still, though his numbers come from the scientific fringe and the coal industry sponsored his big paper, Bezdek's conclusion that more carbon dioxide in the atmosphere could produce net agricultural benefits bears examination. Scientists have long known that CO_2 boosts plant growth, and that fact has been recognized by economists studying the impacts of climate change, including Hope, Nordhaus, and, most enthusiastically, Tol. This kernel of scientific credibility has enabled climate contrarians to spread the gospel of CO_2 fertilization, and this includes contrarians much higher up the food chain than Bezdek.

Take William Happer. A notorious climate denier, Happer is an emeritus physics professor at Princeton whom President Trump appointed to the National Security Council in 2018. Happer used his position to promote climate denial throughout the executive branch. Happer extols the virtues of carbon dioxide and defends it against the "nightmarish police state" that he feels is persecuting this benevolent gas. Here's a sample of his CO_2 remarks over the years: "If plants could vote, they would vote for coal." "If you have a well-designed coal plant, what comes out of the stack of the plant is almost the same thing that comes out of a person's breath." "The demonization of carbon dioxide is just like the demonization of the poor Jews under Hitler." "Demonization of CO_2 and people like me who come to its defense is nothing to be proud of. It really differs little from the Nazi persecution of the Jews, the Soviet extermination of class enemies or ISIL slaughter of infidels." Happer quit the council after serving just over a year.

The reality of carbon dioxide fertilization is more complex than Happer and Bezdek would have it. One problem with much of the science used by Tol and other modelers, including the science regarding carbon dioxide fertilization, stands out immediately; it's old, as Tol himself acknowledges. In 2009, when the IWG was developing the SCC, he coauthored a memo pointing out that FUND "frequently relies on literature that is a decade old or more." When Tol and I spoke, he confirmed that many outdated studies still lurk in FUND and other IAMs. In a workshop discussion on IAMs and the SCC, economist Frank Ackerman came up with an elegantly phrased summation of the problem: "Many of these models carry the fossils of older science."

In 2017, Frances C. Moore et al. published a paper in *Nature Communications* that used the latest science on climate change and agriculture to recalculate the SCCs produced by the Big Three IAMs. The authors found that these models, especially FUND, had significantly underestimated the damage that global warming will do to farming. The study concluded that the SCC would more than double if the IAMs updated their agricultural inputs.

In addition, a great deal of fertilization research has taken place in enclosures—emblematic of the physical and perceptual barriers that separate much of orthodox economics from biophysical reality. Not surprisingly, conditions out in the natural world often produce results that differ from what happens in a lab or a greenhouse, which is why numerous scientists began conducting "free-air CO_2 enrichment" (FACE) experiments. Instead of releasing CO_2 into a chamber full of plants, FACE researchers usually rig a system of pipes that vents carbon dioxide amid plants in open fields. This approach offers a realistic simulation showing how plants likely will react to increased levels of CO_2 under actual growing conditions.

One study of FACE projects examined the reaction of wheat and rice to CO_2 fertilization. Growth of these key global crops did indeed increase, but, as the paper states, "both these grains have shown overall smaller increases than were expected based on earlier enclosure studies." Other studies using FACE data came to similar conclusions, such as one in the *Journal of Experimental Botany* by Andrew Leakey et al. that states: "Overwhelmingly, this [FACE approach] has shown that data from laboratory and chamber experiments systematically overestimate the yields of the major food crops."

This reality check matters because it raises the specter of multitudes of people, present and future, not having enough to eat. In the bland but chilling words of one important FACE study, "If chamber experiments have overestimated the direct effect of increased [CO_2], this would have a major impact on projections of future crop yields." The study goes on to say that the discrepancy between FACE results and chamber results "has wide importance as the chamber values have formed the basis for projecting global and regional food supply, and the stimulation attributed to elevated [CO_2] has commonly been presumed to offset yield losses that would otherwise result from increased stresses." In other words, it would be dangerous to count on a massive fertilization effect to help feed people on a warming planet when that effect will be more modest than massive; whatever minor benefits CO_2 fertilization may temporarily provide will be overwhelmed as temperatures rise.

FACE experiments and other research have also revealed another notable facet about the effect of rising CO_2 levels on plants, one that generally gets overlooked in the debate over the impact of climate change on crop production. In a 2017 commentary for the Heritage Foundation's publication *The Daily Signal*, Republican representative Lamar Smith, then chair of the House Committee on Science, writes that a higher concentration of carbon dioxide in our atmosphere "correlates to a greater volume of food production and better quality food." I've already cited research that refutes his claim of "greater volume," but a growing body of evidence also contradicts his claim about "better quality food." Experiments in plants ranging from oceanic phytoplankton to the world's main food crops indicate that more CO_2 prompts many plants to produce more sugars and fewer vitamins, minerals, and protein. Global warming seems to be slowly turning some of our fruits, vegetables, and grains into junk food.

IT'S COMPLICATED

The sheer complexity of the science that suffuses IAMs is daunting. This complexity stems largely from the intricacy of ecology. After all, ecology comes down to relationships—the vast web of interconnections weaving together organisms and the physical environment, such as the climate. And, like most relationships, they're complicated.

For instance, you might think cod live a simple life, but you'd be wrong. Among other things, it seems that cod have regional accents that can lead to romantic flameouts. Steve Simpson, an associate professor of marine biology and global change at the University of Exeter, has been studying cod vocalizations in UK waters. Simpson reports that these cod communicate by using their swim bladders to make a variety of growls and thumps. But not all cod growl and thump. For example, according to a report in *Science Daily*, American cod "make a staccato, banging, bop, bop, bop sound." This matters, because vocalizations play a key role in cod courtship.

Normally, a cod will breed in the same place its mama and daddy bred, and their mamas and daddies before them, and so on for untold generations. Such geographical isolation makes it more likely that they will have developed distinct vocalizations. As long as they stick close to home, they won't encounter any language barriers when seeking a mate. However, like other fish species in the Northern Hemisphere, cod might be forced by global warming to migrate north in search of cooler waters. If they end up in foreign territories where they don't speak the lingo, their witty banter will go unappreciated, perhaps leading to reproductive failure and downward-trending populations.

Not surprisingly, IAMs don't take the impacts of cod

dialects into account, at least not specifically. The modelers hope to subsume such ground-level effects of climate change through aggregated information. In this case, an IAM might capture any accent-related declines in cod populations by incorporating general data on the productivity of Atlantic fisheries. Then again, such a decline might elude capture. By their nature, IAMs must simplify heaps of complicated inputs, and sometimes that leads to scientific shortcomings, as the modelers themselves recognize.

For a prime example of those shortcomings, we need look no further than the process mentioned above: aggregation. Let's hear from economist Michael Hanemann, a leading voice regarding the perils of aggregating. While making a presentation in 2014 about the then-latest Intergovernmental Panel on Climate Change assessment, Hanemann put up a PowerPoint slide with the heading "Tension: IAMs versus reality." This heading referred to what many scientists thought was a key simplification found in the IAMs: making calculations about climate change impacts based on the average rise in global temperature for a given amount of carbon dioxide emissions. Hanemann felt that using a global average masked large local temperature changes and the subsequent local impacts. Importantly, he told me, when researchers look at local changes, they find "an asymmetric effect that pushes damages up, not down"—meaning that aggregation usually leads IAMs to underestimate, not overestimate, the harm caused by global warming.

You might figure that higher-than-average temperature rises in some places and lower-than-average rises in others would balance out and result in the same amount of overall impacts, but Hanemann would disabuse you of this seductively neat but unsupported conclusion. Studies conducted by Hanemann and others find that the higher local temperature rises in some areas do so much harm that the lower temperature rises at other locales don't offset the higher-than-average places.

An expert in agricultural economics, Hanemann and his coauthors illustrated this asymmetric effect with their research on the impact of climate change on farming in the eastern United States. Using various emission scenarios for the period from 2020 to 2049, they examined the relationship between warming and farm values county by county, which produced exceptionally localized results. They found that if each of these counties warmed by the amount of the estimated average global temperature rise, there would be little influence on crop growth and therefore little influence on farm values. But in the real world, temperatures will not rise uniformly; heat will climb a degree or two or three more in one county than in another. The researchers also found that what caused almost all the heat-related damage to crops was days of extreme heat during the growing season, with "extreme" defined as hotter than 34°C (93°F). Many crops do well up to about 32°C but struggle when subjected to a number of days at 34°C and above.

Hanemann and his coauthors concluded that under various emissions scenarios it appears likely that farms in the relatively cool counties will fare about the same as they would have without climate change while the farms in the hot spots will be hit hard. Depending on which warming scenario they used, the researchers found the overall losses in farm value in the eastern United States would range from 10 to 25 percent. That's not small potatoes. However, if Hanemann and company had simply projected the average global temperature increase onto each of those counties, their study would have concluded that little damage would occur anywhere because this fiction of an average increase would, on paper, push very few of the counties up past the 34°C threshold.

Speaking to the larger point, Hanemann told me, "All

impacts are local." He readily acknowledged the practical difficulty of exploring climate effects at a local scale, especially for IAM developers working on their own or with just one or two others, but he reaffirmed his belief that such research is vital. I said something about it being unfortunate that this microlevel research is so challenging but that I figured it's inevitable given the complexity of our planet. "Yes, it's God's fault," said Hanemann, laughing. "Address your complaint to Him. But one has to get over it and come to grips with it."

MISSING IN ACTION

A few years back I was walking through Cordova, Alaska, bucking a headwind that whipped cold rain against my face. Even during the summer the weather can bite in this little fishing town, a remote dot on the eastern shore of Prince William Sound that can be reached only by boat or plane. The wet, blustery weather has motivated many residents to shun shoe fashion and wear their knee-high rubber boots pretty much all the time, whether on a gillnetter or in a café.

After passing through a ramshackle neighborhood where moose antlers sprouted from weather-beaten garages and long strips of salmon hung on backyard smoking racks, I came to the harbor, where scores of small commercial fishing boats hugged the docks. On the street above the harbor I arrived at my destination: Cordova District Fishermen United, an organization that serves the area's commercial fishers. As I waited to meet with one of the staff, I browsed some of the literature CDFU had laid out for its members. There, amid handouts on fishing gear and the latest regulations, I found a pamphlet warning about the dangers of ocean acidification.

Ocean acidification is exactly what it sounds like: the water in the oceans is becoming more acidic. The level of acidity has risen about 30 percent since the Industrial Revolution, when we humans began pumping out massive amounts of carbon dioxide. By the standards of geologic time, this is an abrupt change, and it's having a significant impact on ecosystems long adapted to a particular water chemistry. For example, acidification disrupts shell development in animals such as corals, shrimp, oysters, and some phytoplankton that help form the base of marine productivity. As a National Oceanic and Atmospheric Administration publication states, "When shelled organisms are at risk, the entire food web may also be at risk." Sadly, the current growth in acidity might be just the beginning; using business-as-usual emissions scenarios, studies estimate that by 2100 the acidity of the oceans' surface waters will have increased by about 150 percent above preindustrial levels.

Note, too, that over the last two centuries the oceans have absorbed much of our civilization's carbon emissions, but that storage capacity seems to diminish with each additional gulp of CO_2; today the oceans are absorbing only about a third of our carbon, and that fraction could continue dropping as the water becomes more acidic. The less carbon the seas sequester, the more carbon goes into the atmosphere, accelerating rising temperatures.

Given the growing harm to fisheries, it makes sense that CDFU would inform its members about ocean acidification. On the other hand, climate change is the main cause of the problem, and frequently residents of rural locales and red states like Alaska don't cotton to talk of climate change, viewing it through a politically partisan and culturally conservative lens. Yet during my visit I found that many people in Cordova, including most of the commercial fishers I talked with, acknowledged and worried about global warming. Perhaps its direct threat to their livelihoods made it harder to dismiss— not to mention that northern climes in general are warming much faster than other regions, and the climate damage in

Alaska is blatant, with storm-battered coastal villages eroding into the sea, wildfires burning north of the Arctic Circle, and melting permafrost causing buildings to tilt.

So even in a faraway place in a red state, many inhabitants know about ocean acidification. Yet this forbidding problem did not figure into the IAMs and the subsequent SCC. Nor do many other problems that scientists—and some science-savvy economists—consider important. In 2014, Peter Howard, an economist with the Cost of Carbon Pollution project, a joint effort affiliated with two environmental groups and the New York University School of Law, published a report called "Omitted Damages." It detailed a number of significant climate-related problems that many IAMs did not include. This chapter's earlier discussion of carbon dioxide fertilization illustrated the kinds of biophysical elements that many scientists think IAMs get at least partly wrong, but the MIAs in Howard's report are examples of climate change effects that IAMs have simply left out, or at least neglected in large part. Admittedly, it's hard to calculate all the damages wrought by global warming, but let's remember attorney Sean Donahue's comment during the *CBD v. NHTSA* case: "The one answer that can't be right is zero."

Some of the omissions arise because IAM developers haven't been working with large, interdisciplinary teams. As Hanemann notes by way of example, with empathy, "Nordhaus [the economist who fathered DICE] doesn't know about forestry. He doesn't know about water supply. He doesn't know about health. You need experts in these fields. You may be a brilliant economist, but you can't understand everything. You can't evaluate the literature on fire, on floods, or whatever." In addition, the limits of scientific knowledge sometimes limit the modelers, as in the case of a relatively new field like ocean acidification that lacks abundant data. And to give them their due, modelers try to improve their IAMs over time by adding to and updating the science.

To better understand the significance of these omissions, let's also consider wildfires, which global warming fuels. As I write this, the sunlight shining through my office window is yellowed by smoke. Major fires are burning to the south of me in the Oregon Coast Range and in Northern California, to the east in the Oregon Cascades, and to the north in the forests of Oregon, Washington, and British Columbia. The smoke has driven people with respiratory issues inside, closed schools, and forced people to choose between sweltering behind closed windows at night and opening windows to let in cool air at the cost of polluting their bedrooms with smoke. However, our local troubles seem trivial compared with the tribulations endured by people closer to the fires, not to mention the grievous harm done to people and places actually in the burn zones.

In 2015, for the first time in history, more than ten million acres in the United States burned and the nation spent a record-setting $1.7 billion on firefighting. But such records don't last long these days. In 2017, the infamous California wine-country fires began chewing through the Napa-Sonoma region, just north of San Francisco. Driven by so-called Diablo winds, the blazes rampaged through forests, fields, and cities, incinerating more than six thousand homes, stores, hotels, restaurants, and other buildings; forcing tens of thousands of people to evacuate; and killing dozens of people. In Santa Rosa and some smaller towns, entire neighborhoods burned to the ground. When the smoke had cleared, 2017 entered the record books as the most destructive wildfire year in California's history. And those fires led to 2017 smashing the 2015 record for the cost of fighting wildfires in the United States, surpassing the $2 billion mark. Not to mention the personal and financial losses people suffered. Insurers in California alone paid out more than $9 billion to people whose homes and businesses burned in wildfires in 2017.

Then came 2018. It turned out to be the new most destructive wildfire year in California's history. Firefighters in the Golden State battled more than eight thousand wildfires, including several massive conflagrations. These blazes killed more than one hundred people and razed some seventeen thousand homes and about seven hundred businesses. The Camp fire caused most of this death and destruction when it rolled over the Northern California town of Paradise, killing eighty-six people and destroying almost fourteen thousand houses and many commercial buildings. According to the international reinsurance company Munich Re, the Camp fire racked up $16.5 billion in costs, making it the world's most expensive natural (well, maybe seminatural) disaster in 2018.

Ominously, the Camp fire happened in November. Wildfires are no longer just a summer phenomenon. Thanks largely to climate change, the fire season in many places has started earlier and finished later than in previous decades. The season in the western states is now about two and a half months longer than it was in the 1970s.

One afternoon in the summer of 2019, my extended family gathered for a reunion in the foothills of the Sierras, about sixty miles south of Paradise. After the requisite overeating, I wandered outside with a couple of relatives who have houses in these fire-prone areas. Surrounded by dry brush and scattered oaks, we got to talking about California's epidemic of wildfires. At one point my two relatives swapped stories regarding their fire insurance woes. They spoke of skyrocketing premiums and some private insurers simply refusing to continue providing coverage.

It turns out my relatives' insurance problems reflect a trend. Californians have experienced a dramatic spike in the number of fire policies that insurers won't renew. The state's insurance commissioner said, "I have heard from many local communities about how not being able to obtain insurance can create a domino effect for the local economy, affecting home sales and property taxes." He added, "Without action to reduce the risk from extreme wildfires and preserve the insurance market we could see communities unraveling."

Wildfires are revving up across the world, not just in California. Research shows that in many areas global warming leads to hotter, drier weather, which in turn leads to bigger fires, longer fire seasons, and more intense fires. Scientists expect this trend to continue and likely accelerate in the future in many places, including much of North America. One study estimates that by the end of the century the average area that burns on this continent each year will be 200 to 550 percent larger than it is today.

Does it matter that some IAMs are missing elements of biophysical reality by omitting some of the social costs of carbon, such as the harm caused by ocean acidification and the additional wildfire damage attributable to climate change? Yes. Are these costs large enough to significantly influence the SCC? Almost certainly.

According to Howard, many of the costs IAMs overlook have not been adequately studied. He thinks the creators of the models would have included many of these slighted costs had pertinent research been readily available. To start correcting this oversight, soon after Howard wrote the "Omitted Damages" report, he wrote a follow-up report called "Flammable Planet."

"Flammable Planet" quantifies climate-induced wildfire costs, though this groundbreaking effort is, as Howard emphasizes, a "rough calculation." He pulled together both broad surveys of wildfire losses and studies of the destruction done by particular fires. He incorporated direct market damages, like the reduction in revenues from timber, grazing, and tourism, as well as the loss of property. Howard also included indirect market damages, such as lost jobs, lower tax revenues, slower economic growth, and reduced property values. Health issues caused by spreading smoke, such as hospitalizations and missed days of work, figured in his calculations, too. Howard also worked in the costs of what he terms "adaptation," such as fire suppression (whose average annual price tag has soared 700 percent in the last two decades), fire prevention, the restoration of burned landscapes, and the evacuation of threatened communities. Finally, à la IAMs, he blended this information with data from climate models to estimate the extra expenses climate change would create under various warming scenarios.

Howard concluded that by 2050 the social cost of climate-induced wildfire in the United States would run from $10 billion to $63 billion a year. For the world, the estimates range from $50 billion to $300 billion annually. Howard sprinkles his work with caveats and stresses the need for more research, but still this study of a single IAM omission suggests the as-yet-unexplored depths of the social cost of carbon and social costs more broadly.

Some of IAMs' scientific fumbles may stem from the different perspectives of scientists and economists. When the interagency working group developed the original federal SCC, the economists on the IWG outnumbered the scientists and dominated the proceedings. At times economists and scientists "were like ships passing in the night," as one observer close to the process said. "It's a cultural problem," says Hanemann. "The people who do the impact work mostly are not economists, and they have a fear and loathing of economics.

They bend over backwards not to include economic metrics. On the other hand, for many economists, if there's a paper that doesn't have economic metrics, it's literally invisible to them. So you have a huge, growing literature on physical impacts that completely bypasses the economists."

Dina Kruger, the former director of the EPA's Climate Change Division, was a member of the IWG when it developed the first SCC. She told me that she brought up the problem of methane, an issue she and some of her colleagues had been studying. A potent and widespread greenhouse gas, methane doesn't stay in the atmosphere as long as carbon dioxide, but while there, it does far more damage. Kruger felt the IWG should also consider developing a social cost of methane. However, when Kruger broached the subject, she encountered the wall between science and economics. "Eventually, as we were talking about the different possibilities, [one of the economists] said, 'Well, we don't really know how to model it. And it wouldn't be right to just multiply by the global warming potential, so I think we ought to just put it on hold for now.' At which point I said something like, 'Well, we know the answer is not zero.'" So, despite its major impacts in the real world, methane didn't get addressed.

On a lighter note, Kruger also recounted a story from a 2009 IWG meeting that illustrates that scientists and economists often simply have different interests. "I'm not a PhD econ, and a lot of this stuff probably goes over my head, but I remember sitting in a meeting and we were talking about damages or something and one of the econs said, 'Do we know whether the curve is concave or convex?' Then all of these economists got completely excited and for like fifteen minutes went back and forth. I'm sitting there going, 'Really?'" On the other hand, an impassioned discussion of cod accents by scientists might have prompted a similarly quizzical response from the econs.

Though at times the IWG economists might have responded quizzically to the science, their time in the working group at least introduced them to the economic implications of biophysical realities. Most orthodox economists, however, have yet to make the acquaintance of biophysical reality in terms of its decisive influence on the economy. Neoclassical economics largely fails to recognize the existence of the natural world, let alone its import.

3

GOING PUBLIC

So there's this guy who loves to fish. For his birthday his kids gave him a $50 gift card to his local outdoors store, so one day he heads down there and starts roaming through aisles brimming with reels, lures, and hooks. But our fisher already has all the basics and soon realizes that the store doesn't have what he wants. No store does. You see, he's eager to improve the angling in his area. What he would really like is an impermeable barrier to keep the toxic heavy metals at the abandoned mine from leaking into his favorite trout stream or regulations to prevent clear-cutting along waterways that get too hot for fish when shorn of their shade. But such fishing improvements will remain elusive, the big ones that got away. The market may enable us consumers to choose from among a hundred brands of fishing rods, but it doesn't give us the option to buy a river healthy enough to fish in.

The market often works efficiently for goods and services that exclusively benefit a particular private party, but it fizzles when it comes to producing or allocating many nonmarket items, especially what economists call public goods. Due to this weakness, that angler can't buy healthy waterways in the marketplace because such public goods don't exist in the marketplace. As with externalities, orthodox economists have long recognized the problem of public goods. And, as with externalities, most policy makers, business leaders, and economists have underestimated or simply ignored the importance of public goods. This renders economically invisible much of what we humans prize most highly. Bringing public goods into full view is one of the main goals of sustainability economics. The SCC is a prime example of an attempt to account for the value of a public good because the market can't do it.

The benefit humanity derives from a stable climate epitomizes what economists define as a public good. (You may have heard some people use the term "climate disruption" instead of "climate change," "disrupted" being the opposite of "stable.") And by "stable," I don't only mean a climate that doesn't hammer us with unnatural extremes of heat, cold, rain, drought, and storms. I also mean a climate that stays much as it has been since the last ice age; a climate like that with which we and our civilizations have evolved; a climate similar to the one that has played a decisive role in where we've chosen to live, what kinds of food we eat, and the kind of work we do. And because climate stability is a public good, "It is impossible to figure out how much climate stability we want through market mechanisms," said Josh Farley, a leading ecological economist and a professor at the University of Vermont.

My fish tale hints at the nature of public goods, but to truly understand how they shake up the conventional wisdom about markets, we need to first understand two qualities that define and separate public goods and private goods: rivalness and excludability.

Let's say you buy an ice cream cone. Once you've eaten it, it's gone and no one else can consume it. That makes an ice cream cone a rival resource (the rivals being other people, say your kids, who always gobble their ice cream cones and then eye yours). Most private goods are rival. Now let's say that you're picnicking on a sunny July afternoon. You forgot to wear sunscreen, but at least your risk of getting skin cancer is decreased by the presence of the ozone layer. Your use of the ozone layer does not make it less useful for anyone else, so it is nonrival. Most public goods are nonrival.

On to excludability. Let's say you own a Segway. And let's pretend, for the sake of argument, that lots of people inexplicably covet your ride. But it's your property, and the fact that you possess it and can exclude others from using it makes it an excludable resource. Most private goods are excludable. Now let's say the solar panels on your roof are busily soaking up power-producing sunlight. But you can't own solar radiation and keep your neighbors from also using it. Sunshine is a nonexcludable resource. Most public goods are nonexcludable.

To kick off our discussion of how the market fumbles public goods, let's consider an inexperienced entrepreneur who makes a pitch to a veteran venture capitalist. This budding young tycoon has discovered that due to the lack of lighthouses, an average of fourteen shipwrecks a year occur on a busy shipping route that winds through a maze of Indonesian islands. He explains to the venture capitalist that with a modest investment the entrepreneur's company could build and operate enough lighthouses to nearly eliminate shipwrecks along that dangerous but time-saving passage, thus providing a service of great value. Thanks to the lighthouses the shipping companies would save hundreds of millions of dollars a year and, says the entrepreneur, it only stands to reason that his lighthouse firm would get a nice slice of that savings. Flush with excitement, the entrepreneur finishes his pitch and eagerly awaits the response of the venture capitalist. She looks at him pityingly and asks, "Who's going to pay you?"

The guiding light from the company's beacons is not rival; one ship's use of it does not use it up so that other ships can't see it. Nor is the service provided by the lighthouses excludable; a private business could not prevent ships from plying these waters or from looking at the lighthouses. Of course, our entrepreneur could ask the ship owners to voluntarily pay for this service. It would be a win-win deal; he would make a decent profit, and the ship owners would pay him much less than what they stood to lose from shipwrecks. But if our entrepreneur had made this voluntary payment plan part of his pitch to the venture capitalist, she would simply have repeated her question: "Who's going to pay you?" Because she knows about free riders.

The principle of free riding is simple; some people are moochers. They're your fellow employees who enjoy the benefits of a clean break room but never clear their moldy cheese sandwiches out of the office fridge. Similarly, the moochers among those ship owners would make the self-serving (orthodox economists would call it "rational") economic choice to not pay for the lighthouses, knowing they could just mooch off a service paid for by others. However—and here's the catch for aspiring free riders—if too many people opt for mooching, then there wouldn't be enough money to enable the entrepreneur to build the lighthouses. And without the lighthouses, the ship owners, including the would-be moochers, would continue losing income to shipwrecks. A solution that would have made everyone better off wouldn't happen. In real life the impulse to free ride frequently causes such lose-lose results when it comes to public goods. (Also, many of us free ride unwittingly, but that's another story, one we'll address later.)

Consider the Pacific bluefin tuna. These sleek apex predators slice through the waters of the open ocean at speeds up to 40 mph. These are not the kind of tuna that ends up in a

can. Chefs, especially sushi chefs in Japan, where most bluefin
are consumed, pay top dollar for these exalted fish. At a Tokyo
fish market auction in 2019, the owner of a chain of Japanese
sushi restaurants paid just over $3 million for a 613-pound
bluefin. That's about $4,900 a pound. No wonder commer-
cial fishers scour the seas for bluefin.

And no wonder overfishing has dragged Pacific bluefin
down toward extinction, reducing the population to an untenable 2 or 3 percent of its original size. A number of countries
have made attempts to restrict the take of bluefin, but too
many free-riding nations and illegal fishers continue to hook
as many of these vanishing tuna as possible. "So if everybody
agrees to restrict the harvest of bluefin tuna," says Farley, the
economics professor, "then you can actually rebuild the stock
and have more bluefin tuna, and because there's more, we can
catch more at a lower price. So everybody wins if everybody
reduces. Collectively, we can get together and set up rules that
limit the harvest of bluefin tuna and other rules that decide
who is entitled to catch part of that limit. Some economists
say that we need private property rights in the market because
people can't act collectively, but, ironically, to create the market, we need to first act collectively. But [regarding the bluefin
tuna] we've chosen not to. In the most recent international
meetings to defend the bluefin tuna, a bunch of people said
that no matter what you do we're going to keep catching
them. So we've essentially chosen to fish them to extinction."

At first glance it seems that the stock of bluefin tuna is the
only public good being squandered by the free-riding folly
described above, but that view is incomplete. The more fundamental public good is the ocean's ability to produce bluefin
tuna. This production stems from the array of ocean ecosystems that nurture this prized fish; the tuna is both a product
of these ecosystems and a contributor to their health. When

ecosystems generate something that benefits us humans, such as the continuous regeneration of bluefin tuna stocks, it's called an ecosystem service.

We're familiar with tuna and other ecosystem products we extract, like trees for lumber and grass for feeding livestock. However, many crucial services operate in the background; like plumbing and wiring, they go unnoticed and unappreciated unless they fail. Consider just a few such services out of the many that a forest provides: the creation of oxygen, the reduction of soil erosion along streams, the absorption of rainfall and subsequent slow release of water later during the dry months, the filtration of pollutants from runoff, the provision of habitat to insects that pollinate crops, the transformation of decaying vegetation into soil, and the moderation of floods.

Forests also render an ecosystem service that plays a major role in mitigating climate change and the calculation of the SCC: carbon sequestration. Trees and forest understory plants store vast amounts of carbon dioxide, making the preservation of forests a vital element of slowing climate change. A 2017 Cornell University study indicates that if we continue the current rate of tropical deforestation until 2100, releasing all that stored-up carbon, it will lead to twice as much warming as scientists had previously thought. The researchers estimate that by the end of this century, even if the world entirely stops emitting greenhouse gases, such deforestation alone would push the planet the rest of the way to 1.5°C above preindustrial temperatures, which many climate scientists consider the point at which we'll be crossing into the danger zone.

Ecosystem services number among the planet's public goods, which means that services like those provided by forests are not accounted for in the marketplace. Does this matter? Does the market's inability to supply and allocate public goods matter enough in a multitrillion-dollar economy to make that inability significant?

NET WORTH

"Everyone in the world depends completely on Earth's ecosystems and the services they provide, such as food, water, disease management, climate regulation, spiritual fulfillment, and aesthetic enjoyment." Thus reads the first sentence in the "Summary for Decision-Makers" of the Millennium Ecosystem Assessment's synthesis report. This UN evaluation of the state of the world's ecosystems took several years, involved nearly 1,400 researchers, and produced clouds of data. Yet that first sentence is perhaps the hefty report's most striking statement, and the most striking idea in that sentence is expressed in the word "completely." Life on earth can't exist without ecosystem services, which makes their value infinite.

We may broadly understand that ecosystem services have infinite value and that we'd better handle them with care, but we need to go further if we want to provide practical guidance for making day-to-day economic decisions and developing policy. In our daily lives we constantly make trade-offs in which ecosystem services are not treated as all-or-nothing propositions. When loggers buzz-saw an acre of forest, we lose much of that acre's trove of ecosystem services, yet even the most zealous forest guardians don't assert that no one should ever cut down another tree. We can afford the trade-off of degrading some of our forests' ecosystem services in order to enjoy the benefits of lumber for building our houses. But how much degradation of services is that lumber worth? How many acres can we afford to log? What's the optimal balance between having lumber and having the forests' ecosystem services?

In search of guidance, various groups of researchers have valued ecosystem services in dollar terms. The most prominent group published a widely cited paper in 2014 called "Changes in the Global Value of Ecosystem Services," by Robert Costanza et al., an update of a famed 1997 study conducted by some of the same researchers. Acutely aware of the immeasurable value of ecosystem services, the authors make clear that by putting a dollar figure on such services they don't mean to treat them like commodities that can be traded in the private market. As the authors put it, "[Ecosystem services'] value in monetary units is an estimate of their benefits to society expressed in units that communicate with a broad audience. This can help to raise awareness of the importance of ecosystem services to society and serve as a powerful and essential communication tool to inform better, more balanced decisions regarding trade-offs with policies that enhance GDP but damage ecosystem services."

Now, on to the dollars. Using 2011 data, the authors estimated the annual value of the world's ecosystem services to be . . . a really big number. More than the approximate wealth of Jeff Bezos and Bill Gates combined. Throw in Warren Buffett and Mark Zuckerberg, and ecosystem services still win, by a lot. What if we included the roughly $200 billion Americans spend each year on fast food? Not enough—ecosystem services come out ahead. Maybe if we added the $16 billion or so that we shell out each year for cosmetic surgery, the $23 billion for pet food, the $77 billion for jewelry, and the $180 billion for apparel? Not even close. Our total remains far short of the value of ecosystem services. It looks like we've got to go really big. How about we throw in the global oil and gas industry's $1.26 trillion in revenues in 2013, when prices were high? Sorry, still a trickle and not a gusher when compared with nature's services. (Besides, oil and gas *are* services bestowed by nature.) Okay, it's desperation time. Let's fat-ten our growing mountain of money with the annual GDPs of Russia, Italy, France, Germany, and Japan, plus America's world-beating GDP of some $20 trillion—altogether that adds about $35 trillion to our total. Surely that's enough to . . . still no.

Okay, it's time to show you the money. Costanza et al. estimate that in 2011 the planet provided us with annual services worth around $125 trillion. Contrast that with the GDP of all the nations of the world in 2011: $75 trillion. And let's not forget that, in truth, ecosystem services provide infinite value for us humans because without them we all would die.

As noted, $125 trillion is a really big number (as is infinity, for that matter), and as such it makes the value of ecosystem services feel abstract, remote from our everyday lives. What brings this issue down to earth for me is a table in the Costanza et al. study that estimates the value of various ecosystems per acre per year. An acre is a bit smaller than an American football field, which is a scale I can picture. For example, an acre of temperate forest, like what I can find near my house, produces $1,270 of ecosystem services annually. Estuaries provide services worth $11,702 per acre. Coral reefs give us the highest value per acre at $142,553. That's a sobering figure in light of the recent reports of corals dying due to warming ocean waters. For instance, about a quarter of Australia's Great Barrier Reef is dead. That's a loss of ecosystem services from an area the size of South Carolina. Using the study's estimates of the value of coral reefs, that comes to nearly $3 trillion. That's $3 trillion a year, every year, unless the reef recovers, and that seems unlikely. Numerous studies indicate that on its current trajectory, climate change will probably kill most of the reef within a few decades, which would mean an annual loss of perhaps $10 trillion in ecosystem services stretching into the indefinite future.

Armed with such dollar-per-acre-per-year estimates, we

can now get specific about the trade-offs involving public goods. Consider the Trump National Doral Miami golf
resort. Sprawled across a landscape that used to be more of
a waterscape, this luxury resort occupies eight hundred acres
of former wetlands. Costanza et al. place a value of $56,728
on each acre of wetlands, which means in its natural state
this property provided ecosystem services worth about $45
million annually. Again, the study's authors caution that we
shouldn't use that figure to decide whether the wetland or
the golf resort contributes more to society, but the number is
informative in grappling with trade-offs.

You can bet that no such grappling occurred when Doris and
Alfred Kaskel ("Doris" and "Alfred," hence the name "Doral")
bought and developed the property in the late 1950s and early
1960s. The Trump Organization bought it and tacked on the
Trump name in 2012. Back in the Kaskels' time, hardly anyone knew about ecosystem services, and people viewed those
wetlands as useless "swampland," the generic derogatory term
applied at the time to pretty much all of the wetlands that
used to cover most of South Florida. The Kaskels paid $20 an
acre—$149 in 2011 dollars, adjusted for inflation. Not quite
the $56,728 per acre value estimated for ecosystem services—
and that's $56,728 per year. But consumers shelling out for
a visit to Trump National Doral Miami don't receive price
signals alerting them to the environmental hit all of us suffer when we lose those wetland services. Higher prices would
signal the cost of this loss and would dampen the demand for
this supply of resort vacationing. Perhaps a round of golf at
Trump National Doral Miami should cost several hundred
thousand dollars instead of several hundred.

Supply and demand simply can't handle the truth of public goods.

Which brings us back to our fisherman, the frustrated fellow we met at the start of this discussion of public goods.

You may recall that he wanted improvements to his local rivers and lakes but couldn't find such goods on the shelves of an outdoors store. When we last saw our fisherman, he was dejectedly roaming the store, but that's not the end of the story.

So what did our disappointed fisherman do? Well, his kids had generously pooled their lawn-mowing money to give him that $50 gift card, so he didn't want to go home empty-handed. Seeing nothing in the store that he really wanted, he finally settled on a talking bass, one of those motion-activated fake trophy fish that blurts out snide comments when someone passes within range of its sensor. He figured the kids would get a kick out of it, so he bought this last-resort item.

With that reluctant decision our fisherman exposes a major flaw in the market economy. He will walk out of the store unsatisfied and will most likely silence the irritating bass with a hammer inside of a week, yet the market will misinterpret his purchase as a signal that a consumer craved a snarky bass above all else that he could have spent that $50 on. The market will have no clue as to what our fisherman actually wanted.

This form of market failure isn't limited to utterly unwanted items like the garrulous fish. For example, someone may have a mild hankering for a new car, but she may have a much stronger yearning for a human-scale town in which she could easily get around by bicycle or on foot. However, she can't go to an urban planning dealership and pick up a compact community, so when she buys a car, she ends up poorly served by the marketplace and her grudging purchase sends a misleading signal to the economy.

The misleading signals created by the market's failure to include public goods throw sand in the gears of a key market mechanism: competition. We all know about the Economics 101 concerns regarding monopolies; if only one company made shoes, we would have few choices, high prices, low quality, and little innovation. Competition among different sectors matters, too. We might have $100 that we could spend on a new pair of shoes, but it's not just a matter of deciding which pair of shoes to buy. We also could spend that $100 on a blender, a chain saw, computer repair, a concert, or a collection of Jar Jar Binks action figures, including one that's a squirter toy. According to orthodox economists, this intertwined universe of competition for consumers' dollars sends the signals to producers that lead to the best allocation of our resources. That would be true, or at least partly true, if all we needed for a good life was footwear and Jar Jar Binks action figures.

However, as we saw above, there's more to life than shoes and squirter toys, with ecosystem services serving as a prime example. Taken as a whole, the goods and services we buy for our private well-being enjoy a monopoly. They don't have to compete with public goods in the marketplace.

Earlier I mentioned that Americans spend about $77 billion a year on jewelry. Some of those glittery gewgaws are sold by QVC on its 24/7 shopping channel, famous for selling stuff to impulse-buying insomniacs at three in the morning. If you've been up late at night channel surfing, perhaps you've run across these pitches, such as the segment in which a couple of hosts fawn over the Joan Rivers Pavé Bluebird of Happiness Brooch. Yours for only $160, this sparkly brooch features a bluebird in flight. Now, when some stubbly guy in his boxers decides that brooch is the perfect Valentine's Day gift for his girlfriend, do you think he's truly signaling that he prefers the Bluebird of Happiness Brooch to 160 dollars' worth of clean water or flood protection?

Okay, so I'm having a little fun with the bluebird brooch. Honestly, I don't mean to belittle people who have bought it, and I realize that jewelry can bring genuine joy. But I do question whether Americans would spend $77 billion a year on jewelry if such private goods had to compete with public goods like clean water and flood protection. I'll bet that our stubbly friend is a sensible guy who spends a lot less on brooches than on food, medicine, housing, and other necessities because they compete for his dollars in the marketplace and he recognizes their greater relative importance. The problem is that nonmarket necessities—like air and water—get left out of that competition.

In 2011, the EPA released a study estimating the costs and benefits of the 1990 amendments to the Clean Air Act (CAA). They figured that in ten years the costs of compliance would be about $65 billion a year, while we Americans would experience annual benefits of about $2 trillion. That's a 30-to-1 benefit-to-cost ratio. Most of the benefits stem from improvements in human health, notably some 230,000 fewer premature deaths per year by 2020. This study only looked at the benefits related to the amendments' reduction of particulate matter and ozone; the benefits would have been even higher if the EPA study had included other pollutants and damages, such as greenhouse gases and the ravages of climate change. These numbers tell us that when it comes to clean air, we are drastically misallocating resources. If an investor could earn $30 for every dollar he invested, he would pour additional money into that deal.

Of course, to address dirty air the market does have its default answer: buy something to protect your own private space, like an indoor air cleaner. Maybe a 600 CFM XPOWER X-3400A Professional 3-Stage HEPA Air Scrubber, $899.00 at Home Depot. (I picked this air cleaner because it seemed like a good one, and maybe a little bit because its name makes it sound like some kind of superhero weapon.) But before buying this or any other air cleaner, you should go to the EPA website's section on indoor air quality and read the article

"Guide to Air Cleaners in the Home." Don't miss the last line: "While air cleaning devices may help to control the levels of airborne allergens, particles, or, in some cases, gaseous pollutants in a home, they may not decrease adverse health effects from indoor air pollutants." This concluding sentence and other information in the article make it clear that the EPA does not consider air cleaners the best solution for dirty air.

In fact, in the first paragraph of the article the EPA identifies the best solution: "Indoor air pollution is among the top five environmental health risks. Usually the best way to address this risk is to control or eliminate the sources of pollutants, and to ventilate a home with clean outdoor air." But in the next sentence the EPA reveals the problem with this approach. "The ventilation method may, however, be limited by weather conditions or undesirable levels of contaminants contained in outdoor air." So you can't deal with your indoor air pollution due to the outdoor air pollution.

Well, like the old political slogan in which a politician promises to put a chicken in every pot, the government could put an XPOWER X-3400A etc. in every household in America—for about $113 billion. Of course, even after this massive expenditure of our tax dollars, we'd still have to live with the shortcomings of air cleaners described by the EPA. Not to mention the shortcomings not mentioned by the EPA, such as the fact that the $113 billion purchase price doesn't cover the externalities associated with the mining, manufacture, transportation, and retailing of air cleaners and their parts or the fact that we'd still be breathing dirty air whenever we ventured outside. Given the relatively low cost of cleaning up much of our nation's outdoor air pollution, plus the extraordinary return on investment when all the benefits are factored in, a collective effort to reduce air pollution seems to easily win the competition when compared with the individual-centered, market approach.

The trade-offs involving clean air have probably been more thoroughly studied than the costs and benefits of any other ecosystem service, yet the value of mitigating climate change has not figured in most of those calculations. That's a gap that the SCC could help fill. The more we understand the worth of ecosystem services and other public goods, the better we can allocate resources.

4

FAT TAILS

Baffin Bay shimmers between Greenland and the northeastern islands of Canada, above the Arctic Circle. If you happened through these frigid, sometimes ice-choked waters you might encounter a pod of narwhals. Inevitably, people dubbed these small whales the "unicorns of the sea" due to the stiletto tusks that protrude as much as ten feet from their heads. Back in the day smooth-talking traders would convince customers that these tusks were actual unicorn horns and sell them for ten times their weight in gold. Narwhals range throughout Arctic waters, not only horizontally but vertically—they dive as deep as a mile in their hunt for cod, squid, halibut, and other prey.

During the last few years some of those Baffin Bay narwhals have been enlisted in another hunt. About a decade ago marine biologists attached sensors to a few of these whales in order to learn more about their habits. When word of the high-tech narwhals reached Ian Fenty, a NASA oceanographer involved in Arctic climate research, he figured some of these wide-ranging beasts could also gather climate change data in the deep waters and remote reaches of the bay. So now the narwhals are capturing information about water temperatures and salinity that will help scientists better understand the relationship between warming waters and the melting of Greenland's ice. In their travels the narwhals also might find evidence of fat tails.

A fat tail is not a sea creature but a statistical concept. The term came into vogue about a decade ago in the financial world. At about the same time fat tails also edged into the IAM and SCC conversation. At first they were interlopers hovering on the margins, but in recent years fat tails have become increasingly influential. Some economists and scientists hope they become transformative.

Whether the subject is math test scores, human heart rates, or preferred thermostat settings, distribution patterns plotted on a graph often resemble a bell—hence the storied bell curve. Most of the numbers cluster toward the middle, where most people and things reside, creating the gently bending line that forms the top of the bell. As the numbers move away from the middle in both directions to depict increasingly uncommon positions, the lines forming the sides of the bell turn downward, sometimes steeply, until they just about touch the bottom of the graph. But often, just before the lines zero out at the bottom, they flatten and stretch almost horizontally for a ways on both the left and right sides of the bell before finally touching bottom. The slender spaces between these almost horizontal lines and the bottom of the graph represent the rare extremes of the people or things being studied. These spaces are called tails. For example, the tail extending along the bottom left of a graph plotting the distribution of preferred thermostat settings might capture the tiny minority of people who prefer settings between 40°F and 50°F. The tail on the bottom right might show the tiny minority who prefer cranking the heat up to somewhere between 90°F and 100°F. The fact that these tails are thin conveys the rarity of these extreme preferences.

In the realm of climate change, tails typically represent unlikely events that would cause major consequences if they were to occur. A few of those consequences could be positive, but far more would be negative, and it would not be hyperbole to label some as calamitous. That's why climate scientists welcome thin tails; their thinness indicates an extremely low probability. But sometimes a bell curve will produce a tail that is thicker than normal, meaning the chance of an outlier event occurring is more likely than normal for a tail (say, a one in 50 chance as opposed to a one in 1,000 chance). This thicker extension of the bell curve is called a fat tail. Climate fat tails are decidedly not welcome.

The makings of a climate fat tail may exist in Greenland, which is one reason Fenty dragooned the narwhals in the hunt for data. As the Arctic's air and water warm, the hulking ice sheet atop the island—about two miles thick in the middle—is melting at an accelerated rate. According to a 2019 study published in the *Proceedings of the National Academy of Sciences*, about five trillion tons of Greenland's ice has melted since 1972—nearly half of it from 2010 to 2018. On July 31, 2019, a record-breaking ten billion tons melted in single day. The next day twelve billion tons melted.

According to recent NASA studies, if all of Greenland's ice melted, sea levels around the world would rise about twenty-four feet. This would lead to many cities going the way of Atlantis, tens if not hundreds of millions of coastal residents being driven from their homes, trillions upon trillions of dollars lost to damages and desperate adaptation efforts, and low-lying regions like South Florida and much of Bangladesh vanishing from the map. People take comfort in the knowledge that a full melt of Greenland's ice very likely won't occur for many centuries, if ever. But people shouldn't get too comfortable; "very likely" is the kind of phrase that suggests the possibility of a fat tail, especially in a fluid situation like Greenland's, in which the scientific knowledge regarding factors like the melt rate has been rapidly evolving. Such uncertainty characterizes many fat tails and looms large in determining the SCC and in sustainability economics in general.

People will feel less comfortable yet when reminded that sea level rise is by no means limited to what happens in Greenland. (Nor limited to melting ice alone—the thermal expansion of warming water molecules adds significantly to rising seas.) Antarctica harbors about nine times as much ice as Greenland, and it, too, is melting, though not as rapidly. Yet. But recent studies of the West Antarctic Ice Sheet reveal increased melting of key coastal glaciers that function as a kind of plug that prevents vast amounts of upslope ice from pushing down to the sea, where the ice would melt. The studies indicate that the melting of those glaciers and a large connected area of ice is unstoppable—a matter of when, not if. The researchers expect the melting to proceed slowly for a long time and then abruptly shift into rapid collapse, causing the ice in that area to disappear within a few decades, adding several feet to sea level rise. Once that plug is gone, the rest of the West Antarctic Ice Sheet is vulnerable. Its melting would raise sea levels by ten to thirteen feet. Depending on how quickly the planet warms, the researchers currently put the onset of rapid collapse at between two hundred and one thousand years from now, but "currently" is another word that conveys uncertainty.

Then there's the much larger East Antarctic Ice Sheet, long thought to be reassuringly stable. But mounting research has eroded some of that reassurance, and we're faced with a low-probability but high-cost event: scientists estimate that the East Antarctic Ice Sheet holds enough water to raise sea level 174 feet. The news from Antarctica grew even grimmer in 2019, when the *Proceedings of the National Academy of Sciences* published an unnerving study about the accelerating pace of the great melt. Between 1979 and 1990, Antarctica was shedding ice at an average rate of forty billion tons a year. Not good, not good at all. But the new study found that since 2009 ice has been melting off the southern continent at the rate of 250 billion tons annually, and researchers expect the ice loss to only get worse.

Nor does melting ice have a monopoly on fat tails. For years Tim Lenton, a professor of climate change and earth system science at the University of Exeter in the United Kingdom, has been a leading voice in sounding the alert about potential climate disasters that are rife with uncertainty but perhaps probable enough to qualify as fat tails. In addition to melting polar ice, he points to the possibility of chronic drought in the Amazon rain forest, which could lead to the dieback of much of this precious ecosystem. He points to indications that climate change might cause larger swings in El Niño, which would probably result in worse flooding and droughts in many regions in and around the Pacific Ocean. He points to the possibility that climate change could disrupt West African monsoons and cause major droughts and the possibility that climate change could disrupt Indian monsoons and cause major floods. He points to the potential loss of the boreal forest, which rings the Arctic, if climate change warms the region beyond the survival threshold of the trees. The unshaded grasslands that would replace the forest would be much drier, possibly resulting in more wildfires emitting more carbon dioxide, which would lead to more global warming, which would lead to more drying, and so forth. He points to another feedback loop that could occur across a huge swath of northeastern Siberia if its permafrost melts, releasing enormous volumes of stored greenhouse gases and triggering a downward spiral. Ominously, in 2019 researchers returned to a study site in the Canadian Arctic and found that the permafrost was melting seventy years earlier than predicted.

Lenton was part of a large international team that published a study in 2018 that took the threat of tipping points to an even higher level. They identified ten major systems in which rising temperatures could trigger nonlinear behavior, including some feedback loops. On its own, each case could wreak havoc, but a few researchers think it's possible that some of these nonlinear effects would influence each other. "These tipping elements can potentially act like a row of dominoes," said coauthor Johan Rockström in a statement from the Stockholm Resilience Centre. "Once one is pushed over, it pushes Earth towards another. It may be very difficult or impossible to stop the whole row of dominoes from tumbling over." If too many dominoes fall it could result in what the researchers ominously refer to as "hothouse earth," a condition in which this chain reaction of runaway feedbacks leads to sharp temperature increases.

Sure, most fat tails will not come to pass; after all, fat tails are by definition low-probability events. But are the odds low enough? In fact, a few of the aforementioned catastrophes might not even qualify as fat tails because the chance that they'll take place may be so great that they don't occupy the tail of the bell curve. Anyway, it's best not to get mired in the taxonomic mud in a fruitless effort to precisely classify what qualifies as a fat tail and what doesn't. Fat tails lie along a continuum ranging from pleasantly plump tails to morbidly obese tails (not real scientific terms). But there is no place on that continuum that we want to be.

THE DISMAL THEOREM

Risk and uncertainty pack a potent one-two punch. Even though we get practice every day as we decide our way through life, it's still hard to make decisions when you're partly in the dark. Maybe you like to indulge in a Krispy Kreme Oreo Cookies and Kreme doughnut once in a while, but you have a heart condition. You have a pretty good idea that you're upping the risk of a heart attack by eating a deep-fried doughnut stuffed with Oreo Cookie and Kreme filling, dipped in dark chocolate icing, and topped with Oreo cookie pieces and white 'n' glossy drizzle. But to what degree are you upping your risk? What if you switched to Krispy Kreme's New York cheesecake doughnut? What if you ate three servings of kale for every doughnut you gobbled? What if you jogged to Krispy Kreme every time you surrendered to your doughnut lust? Questions abound.

As if the uncertainty surrounding problems like the Krispy Kreme–heart attack conundrum weren't enough, it so happens that uncertainty, like doughnuts, comes in different flavors. To kick off our exploration of those varied flavors, let's turn to a famous and wonderful quote from former U.S. secretary of defense Donald Rumsfeld, a man who otherwise did little that would qualify as wonderful; he's best remembered as one of the central perpetrators of the misbegotten war in Iraq. Rumsfeld uttered this quote in what seemed to be an effort to dodge questions at a Pentagon news conference during the run-up to the Iraq invasion. Here is the salient portion: "As we know, there are known knowns; there are things we know we know. We also know there are known unknowns; that is to say we know there are some things we do not know. But there are also unknown unknowns—the ones we don't know we don't know." Not that this quote captures all the gradations and subtleties of uncertainty, but Rumsfeld almost poetically conveys the elusive nature of the matter.

Earlier in our discussions of the SCC we encountered the garden-variety uncertainty found in climate science and climate economics. Much has been learned in both fields, establishing many known knowns, but plenty of known unknowns remain, and by definition, no one knows how many unknown unknowns reside in the shadows. As noted in previous chapters, to varying degrees critics see this uncertainty as a weakness. Some of these critics are fair-minded souls with legitimate concerns.

However, some of these critics are neither fair-minded nor legitimate. Fossil fuel advocates frequently wield even ordinary scientific uncertainty as a club with which to bludgeon scientific results that displease them. Years ago a fisheries biologist told me the exquisite term he uses to describe this gambit: "exploiting the gray." As if to illustrate exploiting the gray for us, early in 2018, then EPA administrator Scott Pruitt made a series of public pronouncements in an effort to turn uncertainty about the details of climate change into doubt about whether climate change would be such a bad thing. During an interview with KSNV, a Nevada television station, the longtime fossil fuel champion said, "We know humans have most flourished during times of, what? Warming trends. So I think there are assumptions made that because the climate is warming, that that necessarily is a bad thing. Do we really know what the ideal surface temperature should be in the year 2100, in the year 2018? That's fairly arrogant for us to think that we know exactly what it should be in 2100." Pruitt went on to say, "There are very important questions around the climate issue that folks really don't get to. And that's one of the reasons why I've talked about having an honest, open, transparent debate about what do we know, what don't we know, so the American people can be informed."

Pruitt and other opponents of climate action aren't so voluble when it comes to the implications of the truly serious flavors of uncertainty that we're exploring in this chapter. On the other hand, many of the scientists and climate economists who don't worry about the run-of-the-mill uncertainty that Pruitt decries are greatly worried by the fat tails that populate climate change. A leading figure among these greatly worried experts was the Harvard economist Martin Weitzman.

"[Extreme high temperatures] would effectively destroy planet Earth as we know it. At a minimum this would trigger mass species extinctions and biosphere ecosystem disintegration matching or exceeding the immense planetary die-offs associated with a handful of such previous geoclimate mega-catastrophes in Earth's history." Disintegration. Die-offs. Mega-catastrophes. Weitzman's characterization of the magnitude of the potential catastrophes represented by climate fat tails evokes an adjective traditionally associated with economics: "dismal," as in the "dismal science." As part of his exploration of climate fat tails, he developed what has come to be known as the "dismal theorem."

To lay the groundwork for this theorem, I'm going to present a bit of basic climate information, some of it derived from *Climate Shock*, the 2015 book written by Weitzman and economist Gernot Wagner, a professor at New York University. We start with the fact that prior to the Industrial Revolution our atmosphere contained about 280 parts per million (ppm) of carbon dioxide. (Note that CO_2 levels undergo minor seasonal fluctuations.) When scientists first precisely measured atmospheric CO_2, in 1958, it stood at about 315 ppm. As of this writing, it has reached as high as 415 ppm. Concentrations have not been this high for hundreds of thousands and perhaps even millions of years.

In an effort to establish mileposts on the road to a warming future, scientists estimate how much hotter the planet will get at various levels of CO_2 concentrations, often using 550 ppm (roughly double preindustrial levels) as a benchmark. Many studies regarding the impacts of climate change estimate what would happen if we hit 550 ppm and stayed there for a long time. This is the approach taken by the Intergovernmental Panel on Climate Change (IPCC), the global body whose influential reports include contributions from hundreds of scientists. The IPCC recently calculated that the likely temperature increase in a 550 ppm world would range from 1.5°C to 4.5°C (though even more recent research suggests that upper bound could be 5.6°C).

You may be thinking that you've already spotted the forbidding uncertainty hiding in that last sentence: an increase that could range as high as 4.5°C. And you'd be partly correct; a rise of 4.5°C would surely cause grievous harm and might detonate some of the catastrophic scenarios I mentioned earlier. Kevin Anderson, deputy director of the United Kingdom's Tyndall Centre for Climate Change Research, writes, "A 4 degree C future is incompatible with any reasonable characterization of an organised, equitable, and civilised global community." But even more forbidding is the IPCC's use of the word "likely," for which the IPCC has a precise definition: events that have at least a 66 percent chance of happening. Weitzman and Wagner employ some sophisticated mathematics—they subtract 66 from 100—to determine that the IPCC figures there's as much as a 34 percent chance that a concentration of 550 ppm will cause a global temperature rise that falls outside the likely range.

Weitzman and Wagner also point out that if the average global temperature change climbs above 4.5°C, it's probably going to climb higher than 4.6°C. They plot the probabilities on a graph, and sure enough, the result is a bell curve with a fat tail. It shows a 10 percent chance of hitting a rise of 4.5°C and then a gradually thinning tail leading out all the way to 10°C before the tail finally tapers down to something close to the 0 percent probability at the bottom of the graph.

Weitzman and Wagner take little solace in the IPCC's determination that a concentration of 550 ppm is very unlikely to push temperatures higher than 6°C above preindustrial levels. "Very unlikely" also has a precise definition for the IPCC: events that have a 0 to 10 percent chance of happening. Zero percent is indeed soothing, but 10 percent certainly does not calm the nerves, not when we are facing the extraordinary consequences of a 6°C rise—or higher.

But wait, there's more. The IPCC bell curve is based on a CO_2 concentration of 550 ppm, but if the world continues on our current emissions trajectory, we will blast past that level. Weitzman and Wagner cite an International Energy Agency (IEA) estimate that includes greenhouse gases besides carbon dioxide, such as methane, and translates their impact into equivalent CO_2 emissions. The IEA figures we'll reach 700 ppm by 2100 unless, as Weitzman and Wagner write, "major emitters take drastic additional steps." A concentration of 700 ppm beefs up that already fat tail on the 550 ppm graph, indicating about a 10 percent chance of temperatures rising 6°C. And this would happen by 2100, not in some distant millennium. Many people born today will be alive in 2100. Many will still be alive in 2101, too, when greenhouse gas concentrations might be 703 ppm, and in 2102, when concentrations might be 706 ppm, and in 2103 . . .

We're far from sure about what will come to pass if we enter the realm of fat tails, but we're confident it won't be pretty. As Weitzman shouted when he took his Paul Revere midnight ride through the streets of academic journals, the mega-catastrophes could be coming, the mega-catastrophes could be coming.

You're probably yearning to dismiss Weitzman and Wagner as a couple of dismal alarmists who, at the sight of sunshine, start muttering about skin cancer. Alas, their views can't be wished away. More and more climate economists and scientists have come to regard fat tails as the most frightening aspect of global warming. To cite just one example, a research group at the University of Cambridge thinks that a temperature rise of 6°C could lead to "the total collapse of the global economy, billions of deaths and the prevention of trillions of future lives," as Simon Beard, one of their associates, wrote in a blog. I picked this particular research group partly because Cambridge is such a renowned university but mainly due to the group's chilling name: the Centre for the Study of Existential Risk.

The specter of existential risk provides an appropriate bridge by which to reach the core of the dismal theorem. You may recall that IAMs pose various climate change scenarios, estimate the damages each scenario would cause, and translate those damages into economic impacts. In order to establish an apples-to-apples comparison, however tenuous, IAMs often express those economic impacts in terms of the percentage of GDP they would cost. This enables policy makers to weigh that estimate against the percentage of GDP that would be lost by spending on climate action. This raises the question at the heart of the dismal theorem: How much of a decrease in GDP is society willing to pay in order to avoid potentially ruinous damages?

Tough question. Basically, the answer depends on the magnitude of those damages, the probability that they will happen, how much of a hit the GDP will take, and how people feel about risk. Many experts think climate IAMs haven't done enough to incorporate the perils of extreme scenarios, though some IAMs have recently improved their handling of fat tails. To demonstrate the shortcomings, Weitzman took the calculations that one mainstay climate IAM used for estimating damages from relatively modest warming and extrapolated to higher temperatures. This extrapolation from the IAM indicated that an increase of 10°C would cause a loss of 19 percent of GDP. Weitzman found this result preposterous. Weitzman would have been the first to say that he didn't know exactly how big the GDP drop would be, but he and many other climate economists and scientists think 10°C would take a way bigger bite than 19 percent.

Now seems like a good time for an encouraging aside before we sink any deeper into dismalness. As I mentioned earlier, and as we'll explore in detail later, sustainability economics offers a hopeful path to stabilizing the climate at a tolerable level of warming while enjoying net gains in quality of life, even if GDP shrinks. So please don't despair. If we act fast and act smart and act fairly, we very likely (my term, not the IPCC's) can pull out of this tailspin. Now, back to dismal.

In general scientists appear to be much more pessimistic than economists—and in this case "pessimistic" seems to be a synonym for "realistic." In 1994, economist and father of DICE William Nordhaus conducted a survey of economists and scientists asking them to estimate the probability of global incomes falling by 25 percent or more under a range of warming scenarios—25 percent being roughly the damage done by the Great Depression. Bear in mind that back in 1994, predictions about climate change were more hopeful than they are today; the knowledge gathered over the last quarter century shows that warming is hitting faster and harder than people in the 1990s expected. Even so, the experts surveyed in 1994 on average estimated that a doubling of CO_2 by about 2050 would result in a 5 percent chance that global incomes would be driven down by 25 percent or more. In an even more rapid warming scenario, the average predicted probability rose to 18 percent.

But as we've seen in other contexts, averages can conceal extreme variation. Under the rapid warming scenario, economists put the probability of a 25 percent or greater GDP drop at between 0.3 and 9 percent, but the scientists put the probability at between 20 and 95 percent. And this was for a scenario in which temperatures climbed maybe 2°C to

4°C, a climb that would almost surely wreak havoc but still be a mere one-megaton explosion compared with the nuclear holocaust of a 10°C rise. In light of all the menacing things we've learned about climate change since 1994, I suspect that nowadays most economists would join Weitzman and company in doubting that 10°C would knock only 19 percent off our GDP.

Along with the magnitude of damages, the probability that those damages will occur is the other key factor we inhabitants of the present must ponder when deciding how much of our GDP we're willing to pay to reduce emissions. If the odds that the global temperature will increase 6°C were one in a million, we wouldn't be willing to pay more than whatever loose change we could find under our couch cushions.

But what if the odds are one in 100 or even one in 20? What if scientists estimate that the odds fall in that 1 to 5 percent range but uncertainty pervades their estimates? Well, says the dismal theorem, when confronted with apocalyptic damages, uncertainty, and a best-guess probability that is not "sufficiently small to give comfort," as Weitzman put it, we should pay up to an infinite amount to protect ourselves. In practical terms, says the theorem, an "infinite" amount means we could pay as much as our entire GDP.

Cue the howls of protest.

The notion of paying an infinite amount led to an uproar among economists. However, in recent years, the fulminating has been fading. Most likely the furor has dimmed partly because those tails are getting fatter and fatter and more and more scientists and economists are sounding the alarm about climate catastrophes. In addition, Weitzman himself downplayed the idea of paying an infinite amount. His formal mathematical model led to that conclusion, but he didn't mean for it to be taken as guidance for action in the real world. As he wrote, "Let us immediately emphasize that which is immediately obvious. The 'dismal theorem' is an absurd result! It cannot be the case that society would pay an infinite amount to abate one unit of carbon." He found the outcry over his infinity result a distraction from his main point: the risk of catastrophe has been largely neglected in climate economics, but it's unnervingly high and we should urgently devote the necessary resources to reducing that risk.

In a 2014 paper Weitzman summarizes his view in a passage that makes delightful use of the family of terms related to "fat tail": "The 'dismal theorem' is best understood as a cautionary tale. A fat tail for rare disasters has the potential to dominate economic calculations like the SCC. Therefore, analysis of a situation that might potentially be catastrophic cannot afford to ignore tail behavior. It is not enough in such situations to look just at measures of central tendency or even just at thin-tailed probability distributions. Ignorance of the potential fatness of an extreme bad tail is not an excuse for ignoring the potential fatness of an extreme bad tail. This warning is the main message of the 'dismal theorem.'" Fortunately, some people have begun to heed that warning.

In saying that fat tails have the "potential to dominate economic calculations like the SCC," Weitzman's words convey a known known. The potential for fat tails to dominate calculations like the SCC certainly exists, but whether that potential will be realized remains uncertain.

The versions of the IAMs the interagency working group used in 2009, when they devised the original federal SCC, mostly overlooked fat tails, though the modelers acknowledged the importance of further exploring them. The working group gave a nod to fat tails but focused almost entirely on higher-probability, noncatastrophic outcomes. Later, as the IAMs' creators continued refining their models, they placed greater emphasis on high-impact, low-probability events. This is especially true of PAGE's Chris Hope, who thinks proper accounting for potential disasters may yield damage estimates that exceed all other projected damages combined.

In a 2011 paper that reflected his evolving views, Hope pondered the possibility of temperatures rising a lot more than expected and a disaster occurring at much lower temperatures than expected. A worst-case scenario involving this "unfortunate combination," as he phrases it with typical British understatement, could result in a cumulative social cost over the next two centuries of $7,000 trillion—yes, that's seven quadrillion dollars. Even Hope's median-case scenario estimates damages at $200 trillion. When the IWG revised the SCC upwards in 2013, much of the increase came from Hope's greater inclusion of fat tails in his revised IAM.

Still, in the eyes of Weitzman and like-minded others, the IWG failed to give fat tails their due. Weitzman said that many IAMs use basic damage functions that "simply cannot register, and therefore will not react to, the possibility of catastrophic climate change" and therefore "almost inevitably [these IAMs] will recommend relatively mild mitigation measures." Such neutered projections drain the urgency from climate policy and support making only slight initial reductions in emissions and then slowly ramping them up to modest levels. Even after some economists injected fat tails into their calculations and came up with SCCs that ranged from several times to dozens of times higher than the government's SCC, the IWG only increased its official figure moderately.

Now that the Trump administration has dissolved the IWG and gutted the SCC, even modest attention to fat tails seems to have disappeared from federal accounting. For the time being, the administration has ordered agencies to rely on a circular from the Office of Management and Budget for all things related to the social cost of carbon. Unfortunately, this circular directs regulators to generally be "risk neutral" and base their analyses on expected, average outcomes, which effectively erases fat tails. Fortunately, contrary to the hear-no-evil, see-no-evil approach of the Trump administration, many climate scientists, economists, and policy makers around the world are starting to come to grips with fat tails, though they have a long way to go.

CATASTROPHE AVERSION

You're given 100 cents. You can keep it all, or you can invest any portion of it in a risky asset. There's a 50 percent chance the investment will fail and you will lose however many cents you invested. Of course, there's also a 50 percent chance the investment will succeed. If it succeeds, you get 2.5 times the amount you invested. How many cents, if any, would you invest?

This scenario is an example of the Gneezy and Potters method—an experimental approach developed by Messieurs Gneezy and Potters to assess risk aversion. The logical move, given the numbers, is to invest all of your 100 cents, but many people choose instead to keep some or all the cents they already clutch safely in their hands. They are said to be risk-averse.

The prevalence of such irrational risk aversion is one of the insights of behavioral economics. Behavioral economics largely grew out of the work of two Israeli psychologists, Daniel Kahneman and Amos Tversky. Starting in the 1970s, these trailblazers reminded the math-obsessed orthodoxy that economics boils down to human behavior. Within a couple of decades behavioral economics had blossomed into a full-grown field that so influenced economics that in 2002 Kahneman won the Nobel Prize in economics, the first non-econ to do so.

But a misapplication of the concept of risk aversion could

undermine efforts to deal with climate fat tails. Yes, in certain circumstances many people are irrationally reluctant to chance losing something they already have, but clearly not all risk avoidance is irrational. For decades numerous representatives of industry and conservative political circles have worked to characterize concerns about many environmental and health risks as irrational. John D. Graham, head of the influential Office of Information and Regulatory Affairs (OIRA) under George W. Bush and a lifelong champion of "rational" regulation (which looks a lot like deregulation), expressed this attitude when he wrote that many such concerns result from the "flustered hypochondria" of consumer and environmental groups. Speaking in favor of Graham's controversial nomination to head OIRA, former senator Phil Gramm (R-TX) made the condescending point even more explicitly: "In reality, what Dr. Graham's opponents object to is rationality." He added, "And they are not rational."

Often rationality is in the eye of the beholder. Consider the example of life insurance, which economist Frank Ackerman cited in his latest book, *Worst-Case Economics: Extreme Events in Climate and Finance*. According to Ackerman, in a given year the odds of dying are about one in 500 for people younger than forty and remain less than one in 100 all the way up to age sixty, yet the majority of Americans buy life insurance. We know that insurance companies make a profit by playing these odds and that their profits represent money we the insured are losing. But some of us also know that our desire to avoid an event that we consider catastrophic (leaving our young children without adequate financial means, for example) overrides our dislike of spending a relatively small amount of money for which we will probably get no return. In light of fat tails, many climate economists and advocates think society should adopt an insurance-like approach to climate change. This view gains even more credence when you consider some key differences between insuring against climate change and insuring against death.

For one thing, standard life insurance is an all-or-nothing situation; you either die and your beneficiaries get a payment, or you don't die (well, for a while) and they get nothing. In contrast, climate damages sprawl along a continuum. Disasters lie in ambush at the far right of the spectrum (they're equivalent to death in the life insurance analogy), but to their left await all sorts of lesser but still painful maladies, such as floods, diminished food production, heat-related illnesses, and water shortages. The resources we put into lowering the risk of climate catastrophe also reduce those noncatastrophic damages. It's like buying life insurance and additionally getting health insurance for free.

Insuring against climate change also differs from insuring against death because it entails greater uncertainty. You as an individual may unexpectedly get hit by the proverbial bus at any time, but as a group our deaths form predictable patterns; that's what allows insurance companies to develop actuarial tables. The probabilities of climate fat tails are more difficult to calculate. In *Worst-Case Economics,* Ackerman notes that the paucity of information about previous periods of climate change deepens the uncertainty. Besides, adds Ackerman, "Climate risks are dangerously uncertain because the future will not resemble the past. Temperatures and greenhouse gas concentrations are moving well beyond the limits of historical experience, making it hard to predict the outcomes, and particularly hard to rule out the worst cases."

So how do we make the best possible decisions about handling the risks associated with climate change? Specifically, in the context of IAMs and the SCC, how much of our present-generation GDP should we spend to protect future generations?

Oddly, I think the answer is clear. Not the exact answer,

but a ballpark idea that would be enough for us to work with—if we decide to do the rational thing. (Rational in the eyes of this beholder, anyway.)

The answer comes into focus when we look at what we stand to lose and what we stand to gain if we don't spend enough GDP to reduce greenhouse gas emissions to safe levels. We've already talked in detail about what we stand to lose; we will almost certainly suffer awful but not earth-shattering damage, and we will face a substantial risk from fat tail events that might ravage entire regions or even end civilization as we know it.

The other side of the ledger shows what we stand to gain by shortchanging climate action—that is, the amount of money we'd have to spend on such things as solar arrays, electric vehicles, reconfiguring cities, and insulating homes. In 2017 the International Energy Agency (IEA) and the International Renewable Energy Agency (IRENA) issued a major report that estimates how much it would cost the world to make sufficient reductions in greenhouse gases. The report defined "sufficient" as meeting the goals of the Paris climate accord, namely keeping the global temperature rise to "well below" 2°C.

IEA/IRENA came up with a gross estimate that the world would need to spend about $145 trillion during the thirty-five years between 2015 and 2050. Take our current annual global GDP (a.k.a. GWP, or gross world product) of about $75 trillion, multiply it by thirty-five years, do the division, and you get a yearly cost of about 5.5 percent of GWP. That would sting, but it's doable. Fortunately, we won't have to do it. The 5.5 percent figure is bogus because my crude extrapolation of future GWP suffers from fatal flaws. Most notably, it overlooks the benefits of switching to a clean energy economy.

The report finds that the switch likely would boost jobs and the economy enough to produce a net gain in GWP of $19 trillion by 2050, creating an annual GWP by that year that would be about 1 percent higher than if we continue on a business-as-usual emissions pathway. The report also mentions the value of reducing some of the social costs of global warming, noting that "improvements in human welfare, including economic, social and environmental aspects, will generate benefits far beyond those captured by GDP." IEA/ IRENA conclude that money spent on making the transition to a clean energy economy would produce enormous net benefits, saving between two and six times more than the costs of making the necessary emission reductions.

Naturally, uncertainty impinges on these calculations, as IEA, IRENA, and most everyone else in this field readily acknowledge. Though it is likely that any major surprises would be positive, such as developing new technologies to more cheaply reduce emissions, we know from past discussion that "likely" is not the same as "certain." Maybe the anticipated dividends from a transition to clean energy would fall short of expectations. It's even possible that net benefits would not materialize at all and society would suffer moderate net costs, but there is no chance, none, that the cost of reducing emissions would drag society down into a *Mad Max* dystopia. In terms of the magnitude of the damage, the risks associated with spending on emission reductions are trivial compared with the risks we'll face if we don't sufficiently reduce emissions.

PASCAL'S WAGER

God is, or He is not . . . Let us weigh the gain and the loss in wagering that God is . . . If you gain, you gain all; if you lose, you lose nothing. Wager, then, without hesitation that He is . . . [T]here is an eternity of life and happiness. And this being so, if there were an infinity of chances, of which one only would be for you, you would still be right in wagering one to win two, and you would act stupidly, being obliged to play, by refusing to stake one life against three at a game in which out of an infinity of chances there is one for you, if there were an infinity of an infinitely happy life to gain.

But there is here an infinity of an infinitely happy life to gain, a chance of gain against a finite number of chances of loss, and what you stake is finite. It is all divided; wherever the infinite is and there is not an infinity of chances of loss against that of gain, there is no time to hesitate, you must give all.

This quotation continues in the same vein and, trust me, it doesn't get any clearer. I wouldn't expect anyone to make sense of these arguments taken out of context. Frankly, it's a slog even taken in the context of the dense work from which it comes. Blaise Pascal, a French philosopher, physicist, mathematician, and Catholic theologian, wrote these words in the seventeenth century. In what scholars widely consider a milestone in decision and probability theory, he argued that a belief in God is the only rational choice in light of the risks and rewards. In his reasoning, not to mention his frequent use of the concept of infinity, you'll hear murmurs of Weitzman's dismal theorem.

To carve Pascal's wager down to the bone, as best I can, it posits that God may or may not exist. If He does exist and you believe in Him, you'll spend the rest of eternity in heaven. If He does exist and you *don't* believe in Him, you'll spend the rest of eternity in hell. If He does not exist and you believe in Him and therefore curtail your sinning, you lose a bit of earthly pleasure. If He does not exist and you don't believe in Him and therefore proceed with your sinning, you gain a bit of earthly pleasure. Considering the infinite upside of believing in Him if He does exist, the infinite downside of *not* believing in Him if He does exist, and the relatively minor pros and cons involved if He doesn't exist, Pascal recommended believing in Him. (Pascal didn't seem to prize sincerity, or maybe he figured that He did not.) Pascal waved away concerns over precise probabilities because when you're talking about taking a chance that entails infinite upsides and, particularly, infinite downsides, such as eternal damnation, the only rational choice is the better-safe-than-very-very-sorry option.

"Better safe than sorry" expresses the essence of the precautionary principle, a key element of sustainability economics that should sway the SCC and climate economics. Note that this is not the cower-in-your-underground-bunker-eating-Spam-because-the-sky-is-falling principle. Advocates of the precautionary principle are not the flustered hypochondriacs scorned by John Graham. They do not fearfully demand that the EPA declare your neighbor's backyard barbeque pit a Superfund site because he spilled a can of lighter fluid. As Frank Ackerman liked to point out, people buy fire insurance, but they don't buy meteorite insurance; most of us don't get too carried away with our concerns about risk. The precautionary principle simply calls for erring a reasonable amount on the side of caution when dealing with uncertain matters, particularly those that could have a big negative impact. Going to hell forever, for example. Or global warming so extreme that we end up stranded in hell on earth.

In addition to having the potential to cause widespread damage, the global warming threats that make the precautionary principle list tend to share certain other traits. Irreversibility comes to mind. The meltdown under way in Greenland provides a menacing example. As I reported earlier in this chapter, the ice sheet atop Greenland is leaking water at a rapidly accelerating rate, contributing significantly to sea level rise. But I didn't mention that numerous studies have discovered that climate change may soon push global temperatures over a line from which Greenland can't recover. After crossing that line, the unwavering physics of heat and ice would inexorably melt the entire ice sheet regardless of how much we reduced emissions.

In some scientific scenarios we've already passed the point of no return. In others we can still stave off most of the melt if we act quickly and strongly. An influential 2012 study in *Nature Climate Change* calculated that a temperature rise of 0.8°C to 3.2°C above preindustrial levels would tip Greenland's melt into the irreversible zone. The researchers' best estimate for when this tipping point would occur was 1.6°C. Well, we've already exceeded 0.8°C, and on our current emissions trajectory, we'll pass 1.6°C in a few decades. Some observers console themselves with the knowledge that it'd probably take a thousand years for Greenland's ice sheet to vanish and boost sea levels by more than twenty feet. But bear in mind that the ice will have been melting all along and proportionately raising sea levels all along; maybe the Greenland melt will cause sea level to rise two feet by the end of this century. As there's no fix for such irreversible situations and the damage would be essentially permanent, invoking the precautionary principle seems like the smart play.

Long lag times provide yet another reason to treat climate change with precautionary kid gloves. When we emit carbon dioxide, it doesn't immediately raise the planet's temperature. Once loosed, carbon can take decades to exert its full impact. This means we've already committed the planet to a fair bit of warming that we haven't yet experienced. It's like chugging a 500 mL bottle of Spirytus vodka and thinking that the 192 proof liquor won't make you drunk just because you haven't instantly felt the effects. I recommend that you use the lag time to get to the bathroom—or maybe the ER.

Downward-spiraling feedback loops rank high among the

traits that call for caution. For instance, our warming climate melts permafrost, which releases stored greenhouse gases, which revs up global warming, which melts more permafrost, and so on. Such feedback loops sometimes lead to irreversible changes, so they can be doubly troubling.

Feedback loops number among the climate processes that can trigger rapid, even abrupt change, which brings up another characteristic of climate change worthy of precaution: tipping points. Nature often reacts to excessive stress in a nonlinear fashion. Consider coral reefs. Since the Industrial Revolution got cooking, they've been feeling the heat as ocean temperatures and acidification gradually increased, but now marine scientists widely agree that global warming has pushed corals to the brink of an abrupt (and perhaps irreversible) decline that could lead to their extinction. In sad fact, some researchers think corals have already been shoved over the brink and most will perish within a few decades.

Corals may be a harbinger of a tippy future. A 2015 paper in *Proceedings of the National Academy of Sciences* searched all the earth system models used in the latest IPCC report for tipping points. Though leavened with caveats about the possible gaps between models and reality, the paper's findings still can make your pulse race. The authors found evidence of thirty-seven "forced regional abrupt changes in the ocean, sea ice, snow cover, permafrost, and terrestrial biosphere that arise after a certain global temperature increase. Eighteen out of 37 events occur for global warming levels of less than 2°, a threshold sometimes presented as a safe limit."

Every time the IPCC examines tipping points, the risks darken. In 2019 a group of climate researchers charted the changes from the 2001 IPCC report to the 2018 report. In 2001 the scientific community calculated that the average global temperature would have to rise around 4°C to create a moderate risk of triggering tipping points and 5°C or more

to create a high probability. But with each report that temperature dropped. By 2018 the IPCC figured a rise of around
3°C would create a high risk and 1°C to 2°C a moderate risk.
Remember that we've already hit 1°C. Given the trend, I can't
say I'm looking forward to seeing the next IPCC report.

Irreversible consequences. Lag times. Feedback loops. Tipping points. All these qualities signal that we should embrace
the precautionary principle in dealing with the fat-tailed
uncertainty that characterizes climate change. So how do we
embrace it?

First, the precautionary principle exhorts us to giddyup. If
we don't fix the leaky pipe now, we'll be dealing with a flooded
basement later, so speed is of the essence.

Second, the precautionary principle tells us . . . well,
nothing really. It urges us to err on the side of caution, but
it doesn't say how to do that. The above IPCC quote continues, "The question of timing and extent of mitigation and/or
adaptation policies remains unquantified by the precautionary principle."

But we already know what to do, at least in broad strokes.
Our earlier discussion of the dismal theorem makes it clear
that we stand to lose far more through inaction than we stand
to lose through action. Besides, as the IEA/IRENA report
indicated, climate action probably will produce a net gain
for society. So we'd best keep the global temperature rise well
below 2°C. If by doing so we end up erring on the side of caution by spending a little more money than necessary a little
sooner than necessary, well, that's a minor error. Much better
than spending too little too late. It would be irrational to take
the risk.

5

THE ONCE AND FUTURE KIDS

Levi didn't look like a threat to the United States government and a trillion-dollar industry. His curly, untamed hair and turquoise shirt said playful, not powerful. His ready smile and guileless countenance added to his aura of harmlessness. Most disarming of all, Levi Draheim was only eight years old.

It was March 9, 2016, and young Draheim was sitting on one of the wooden pews in the U.S. district courthouse in Eugene, Oregon. His feet dangled in the air because his legs weren't long enough to reach the floor. The courtroom was literally packed; a few minutes earlier the clerk of the court had walked around asking people to squeeze together on the long benches to accommodate as many onlookers as possible. Even so, lots of people had to troop to overflow rooms to watch the proceedings on closed-circuit TV. The hearing also was being streamed via video feed to a thronged courtroom in Portland.

Though the hearing was scheduled to begin at ten a.m., I had arrived three hours earlier to make sure I wouldn't be swept away in the overflow. Dozens of people had gotten there before me, most of them elementary school students from a school more than a hundred miles to the north. Soon a crew from a local TV station showed up and began interviewing some of the kids. Several adults held up a twenty-foot-long banner whose bright orange letters read "Our Future Is a Constitutional Right." Over the next couple of hours, more and more people arrived, bundled up in their heavy-weather gear on this wet, cold spring morning. By the time the security guards opened the outer doors, the queue stretched all the way from the entrance down the courthouse steps to the sidewalk below.

After passing through security, we moved upstairs to the hallway outside the courtroom, where we waited another hour or so. The scene during the wait was lively, perhaps because kids made up about half of the two hundred or so people in the crowd. But loud as the hubbub was, the noise level spiked when the plaintiffs arrived and made their way through the supportive crowd. A wave of applause swept them along the hallway and crested as they neared the courtroom doors.

As the plaintiffs came into view, two facts immediately stood out. One, there were lots of them—twenty-one altogether. Two, they were a bunch of kids and teenagers, ages eight (that would be Draheim) to nineteen. Yet despite their youth, they were suing the federal government over climate change. Activists Bill McKibben and Naomi Klein called this "the most important lawsuit on the planet right now."

The overarching issue raised by these youth decisively influences the results of IAMs and forms a cornerstone of sustainability economics. I'm talking about the other E besides environment and economy that goes into the Three E's: equity.

WHO GETS WHAT?

Discussions of equity often begin with some jaw-dropping statistic regarding wealth inequality, especially in the United

States, one of the most unequal of the developed nations (twenty-third out of twenty-nine, according to the World Economic Forum's 2018 report). For example, a recent study found that the three richest Americans together possess more wealth than do all the 165 million Americans who occupy the bottom half of the wealth chart put together.

Typically we frame equity as a moral issue, and it is. But it also has a vital practical dimension, according to considerable research, notably expressed in the groundbreaking book *The Spirit Level: Why Greater Equality Makes Societies Stronger*, by Richard Wilkinson and Kate Pickett. The authors studied data regarding many of the leading indicators of misery around the globe, such as poor health, violence, drug addiction, loss of community life, mental illness, and imprisonment. Unsurprisingly, many impoverished societies suffer from such problems. Surprisingly, so do many rich societies that exhibit significant inequality—and the greater the inequality, the greater the problems. The authors found that those social afflictions are two to ten times more common in more unequal nations.

Orthodox economics takes an agnostic position on equity. Neoclassicists emphasize efficient allocation of resources with the goal of expanding the GDP. Distribution—who gets what—does not enter into the equation, literally. As long as the total amount of goods and services rises, victory will be declared, even if more and more of that wealth ends up in the hands of the 1 percent while less and less reaches the 99 percent.

Free enterprise believers may personally hope for a more equitable world, but they think distribution should be left to the market. In this view do-gooders trying to inject fairness into the economy only mess up the market's genius for yielding the greatest good. Let's listen once more to our singing friends from chapter one, the Milton Friedman Choir. In the musical paean to free markets, Friedman's choir croons that corporations—and, by extension, markets—have "no social duty," "are amoral," and that "corporate conscience is impossible." This vision of the market as a machine operating apart from human values, such as equity, provides one of the main justifications for not wanting to redistribute wealth, notably in the form of taxes that shift money from the rich to the poor and middle class. Of course, some opponents of redistribution don't even know the term "neoclassical economics" and simply want to possess as much money as possible; their disdain for redistribution is self-serving, not philosophical. Redistribution is indeed a key issue, but the mainstream's narrow conception of the matter omits the view that equity involves much more than redistribution via taxes.

Sustainability economists offer a brighter, more ambitious vision. They see equity as a vital social value and believe that an economy should be designed to foster such desirable social values, not just aimlessly crank out material wealth and private goods. Sustainability economists didn't arrive at this position through complex equations or esoteric laws of economics. They just made a value judgment that pursuing equity was the right thing to do.

The tension between those who see climate change through the lens of values and those who don't erupted in the fall of 2015 when Pope Francis issued his famous encyclical on climate and the environment. His nearly book-length papal message called on Roman Catholics—and all inhabitants of the planet—to "care for our common home," as the subtitle of the encyclical puts it. Pope Francis frames this as a moral imperative and brings up the disproportionate suffering of the poor and the disenfranchised caused by environmental degradation. At one point he writes, "Climate change is a global problem with grave implications: environmental, social, economic, political and for the distribution of goods.

It represents one of the principal challenges facing humanity in our day. Its worst impact will probably be felt by developing countries in coming decades. Many of the poor live in areas particularly affected by phenomena related to warming."

Pope Francis's encyclical came out during the American presidential primary campaign, and it was met with disapproval from the Republican candidates, all of whom firmly embraced free market beliefs. Jeb Bush, who is Catholic, said, "I hope I'm not going to get castigated for saying this by my priest back home, but I don't get economic policy from my bishops or my cardinal or from my pope." He continued, "I think religion ought to be about making us better as people and less about things that end up getting in the political realm." Rick Santorum, also Catholic and famously devout, said, "The church has gotten it wrong a few times on science, and I think we probably are better off leaving science to the scientists and focusing on what we're good at, which is theology and morality." He added that climate change is "really outside the scope of what the church's main message is, that we're better off sticking to things that are really the core teachings of the church as opposed to getting involved with every other kind of issue that happens to be popular at the time."

I'm going to skip over the glaring irony of climate-change-denying Santorum challenging the church on climate science and get to my main point; he and Bush both asserted that the pope should stay in his lane of morality and not stray into economic policy, science, and politics. This parallels the laissez-faire stance that moral values, such as equity, ought to stay out of the market. A sustainability economist and the pope would disagree, but apparently Bush and Santorum found both fallible.

Concern about equity usually focuses on the present. That's understandable given that most of us can look around the world and around our neighborhoods and see too many people squeezed by the economy. Later we'll see what sustainability economics has to say about equity in the here and now. But the future always becomes the present, which makes equity an intergenerational question, too. How will we distribute resources between those of us who are adults now and the next generation, or the generation after that, or the generations of the twenty-second century and beyond? This profound matter often gets overlooked, but it had gotten the full attention of the kids who sat in that federal courtroom in Eugene, Oregon, in 2016.

DAVID V. GOLIATH

On August 12, 2015, attorneys for the twenty-one young plaintiffs filed an unprecedented lawsuit against the U.S. government. Here's some of what the lead plaintiff, Kelsey Juliana, said in a statement accompanying the announcement of the lawsuit: "Our nation's top climate scientists . . . have found that the present CO_2 level is already in the danger zone and leading to devastating disruptions of planetary systems. The current practices and policies of our federal government include sustained exploitation and consumption of fossil fuels. We brought this case because the government needs to immediately and aggressively reduce carbon emissions, and stop promoting fossil fuels, which force our nation's climate system toward irreversible impacts." She finished by adding, "If the government continues to delay urgent annual emissions reductions, my generation's wellbeing will be inexcusably put at risk."

Juliana's statement brings up the most prominent of the assertions made by the kids: the federal government is violating their constitutional right to life, liberty, and property and to the pursuit of happiness on a planet that's not unduly degraded by climate change. To say the least, this is a bold claim whose far-reaching implications evoke momentous predecessors, such as the constitutional claims regarding civil rights and LGBTQ rights.

Equally bold and far-reaching are the remedies the youth seek. They're asking the court to order the government to stop enabling the production and use of fossil fuels and instead phase them out rapidly. Specifically, they ask the court to instruct the government to do its part in the global effort to achieve an atmospheric concentration of carbon dioxide of no more than 350 parts per million (ppm) by 2100. Though many scientists have recommended stabilizing the climate at 350 ppm or below, that would require a huge effort. Given that the CO_2 concentration has already reached 415 ppm, the United States and other nations would most likely have to make a rapid transition to clean energy while also removing some CO_2 from the atmosphere. Highly esteemed climate scientists associated with the kids have developed what they see as a realistic pathway for such CO_2 reductions, but it certainly wouldn't be easy to follow that path.

Incidentally, if you look closely at the official complaint submitted to the court, you'll notice that there's a twenty-second plaintiff in the case whose name lurks at the bottom of the long list of youth. When minors appear on a complaint, their full names are withheld. For example, Levi Draheim is listed as "Levi D, through his Guardian Leigh-Ann Draheim" (his mother). Well, at the bottom of the list is the phrase "Future Generations, through their Guardian Dr. James Hansen." I suppose there could be an actual person stuck with "Future Generations" as a name—remember, there's a Moon Unit Zappa out there—but in this case the plaintiff is not some cruelly christened person but all future generations of human beings. And the guardian, Dr. James Hansen (whose granddaughter is one of the youth plaintiffs), is probably the most renowned climate scientist on the planet. Despite his prominence, however, the kids remain the heart and soul of this litigation.

Each of these twenty-one young people has a story, but for the sake of brevity I'll sketch the backgrounds of just a few of them. About half come from Oregon, and the others are scattered all over America. I'll mention their ages as of the time of this writing, but bear in mind that they started their long trek through the courts in 2015, so when they began this legal journey, they were considerably younger than the ages given.

Officially the case is known as *Juliana v. United States*, after lead plaintiff Kelsey Juliana, so let's begin with her. Having turned twenty-three, she is the elder stateswoman among the youth plaintiffs. An energetic native of Eugene, she walked from Nebraska to Washington, D.C., in 2014, as part of the Great March for Climate Action. Currently she works with a nonprofit organization to bring information about sustainability to schools, attends the University of Oregon, and works at a local Eugene bakery. A passionate and articulate advocate, she often gets called upon to speak with the media; her appearances have ranged from CNN to *Teen Vogue* to *Scientific American* to *60 Minutes*. I get the feeling she's something of a leader for the younger kids, too. Just before the March hearing began, I saw her kneeling beside a couple of the little ones, seeming to calm them while perhaps giving them a bit of last-minute coaching on courtroom behavior.

Like Juliana, nineteen-year-old Xiuhtezcatl Martinez of Boulder, Colorado, often speaks publicly about the lawsuit and global warming. He has sounded warnings about climate change in *National Geographic* and *Rolling Stone*, and on PBS and CNN. He has been advocating for climate action and the environment since he was six and has spoken before many government bodies, including giving three speeches to the United Nations. He also was the youngest member on a youth council that advised President Obama. Martinez works as the youth director for an international nonprofit that engages young people around the world to advocate for that world.

Naturally, we should talk a bit more about Levi Draheim, whom you've already met, the youngest of the twenty-one plaintiffs. For all of them climate change is personal and affects their daily lives, and that's most definitely true for Draheim because he lives on a barrier island along Florida's Atlantic coast. Draheim worries that the beaches, which are his backyard and playground, may soon disappear. Located just a few feet above sea level, his community recently undertook a study to determine its vulnerability to sea level rise. The results indicate that Draheim's house will be submerged in a few decades and the whole island will be underwater by the end of the century. In September 2017, Draheim got a preview of things to come when Hurricane Irma thrashed the area and, a week later, torrential rains inundated parts of his town, flooding his house and forcing him and his family to evacuate. They were able to return shortly, but his school was totaled, and for a while he had to be homeschooled.

Sea level rise and stormy seas also threaten the home of Mani Wanji "Journey" Zephier. They don't threaten his birthplace—Zephier is a member of the Yankton Sioux Nation and was born in South Dakota—but they are hammering the small town of Kapaa, his home since 2009, when his family moved to the Hawaiian island of Kaua'i. Like Draheim's barrier island, Kapaa is being overtaken by the ocean and is expected to be mostly submerged by the end of the century. Zephier says he already sees considerable harm from climate change. Nearby coral reefs are dying, beaches are eroding, more intense storms bring floods, and sometimes the island, which used to be one of the rainiest places on the planet, is parched by drought. To fight back, Zephier serves as a leader of Rising Youth for a Sustainable Earth and as a youth ambassador for the Center for Native American Youth.

These thumbnail profiles give you a glimpse of who these twenty-one kids are. And there are more where they came from; Our Children's Trust, the Oregon nonprofit leading the kids' campaign, has initiated such youth-led lawsuits and related legal actions in many states. Some have been dismissed by the courts, a few have managed partial wins, and others are still grinding through the process. They're all important to push the idea forward, says Julia Olson, but the federal case "is definitely the core of the effort." Olson serves as the executive director and chief legal counsel for Our Children's Trust and is the lead attorney arguing for the plaintiffs in the federal lawsuit. After some fifteen years of practicing environmental law, Olson became a mother and, she says, having a child focused her more on the future. That led her to focus on climate change, which she considers the greatest threat to the coming generations. And that in turn led her to found Our Children's Trust and to conceive of the youth lawsuits.

Before we proceed with *Juliana v. United States*, I should mention one other thing. Shortly before the hearing, I talked with and read commentary by numerous lawyers familiar with the case. They ranged across the political spectrum and included some people who strongly support climate action. These attorneys probably don't agree on much, but they all agreed on this: the kids have almost no chance of winning.

The March 9, 2016, hearing in Eugene was the first courtroom proceeding involving the lawsuit. It posed a major, likely insurmountable hurdle that the kids would have to clear to continue toward a full-fledged trial. The U.S. government—the defendant—had filed a motion to dismiss the kids' lawsuit, and the hearing gave the government and the kids the chance to present their arguments about that motion before Magistrate Judge Thomas Coffin. And the government was no longer defending alone. A couple of months earlier, power-house reinforcements had arrived in the form of the American Petroleum Institute, the National Association of Manufacturers, and the American Fuel and Petrochemical Manufacturers. These trade associations represent the bulk of the nation's fossil fuel industry and many of the world's largest fossil fuel companies. Per their request, Coffin had named them official intervenors for the defense, which enabled them to submit documents and appear in court alongside the government. Apparently, the fossil fuel industry was taking no chances with this lawsuit, even if almost all legal observers regarded it as quixotic and nearly certain to fail.

The kids didn't have any intervenors, but numerous allies did file amicus curiae ("friend of the court") briefs in support. For example, the day after the fossil fuel industry entered the case, the Global Catholic Climate Movement (GCCM) and the Leadership Council of Women Religious filed amicus briefs. This once again connects the climate kids to Pope Francis, who had issued his climate encyclical, *Laudato Si*, just months earlier. The GCCM represents some 250 Catholic organizations and individuals, including Catholic bishops and the pope. "As an organization inspired by the principles of 'Laudato Si,' the Global Catholic Climate Movement welcomes the opportunity to support the young plaintiffs," said Tomás Insua, global coordinator with the GCCM. "By supporting this initiative, we join our voices with the young plaintiffs who are calling for climate justice and the protection of the atmosphere for generations to come." Since then many other organizations have submitted amicus briefs on behalf of the youth, including the League of Women Voters, Interfaith Moral Action on Climate, the Center for International Environmental Law, the Sierra Club, Eco-Justice Ministries, the Sunrise Movement, the Union of Concerned Scientists, Earthjustice, and Defenders of Wildlife.

The defense led off the March 9 hearing. Speaking for the government, attorney Sean Duffy did not challenge the kids' claims regarding the science of climate change or its impacts on the planet. Nor did industry lawyer Quin Sorenson, though in pre-hearing documents the industry intervenors sprinkled their arguments with droplets of climate skepticism. Such issues would arise if the case ever went to trial, but deciding whether to dismiss the case involved other matters, matters on which the government and the industry intervenors largely agreed. Duffy got right to one of their main assertions: climate change should be addressed by Congress and the executive branch, not by the courts. "No court has done so," he said, noting the lack of precedents. He then cited various cases and legal theories in support of his claim that courts are not the place to raise generalized grievances regarding government policy.

When it was Olson's turn to stand before Coffin, she stated that the kids had gone to court because the legislative and executive branches had failed in their duty to protect the children and future generations. To illustrate this, she asked an assistant to put up a chart that she said would show what the government had done over the years to address climate change, but the assistant accidentally flipped to a blank page, which was inadvertently appropriate. That got a big laugh from the courtroom. But the chart, once produced, did not elicit any chuckles. It showed that the U.S. government had known about the hazards of global warming for more than fifty years and had done little about it. For example, way back in the 1960s, a White House report and a memo from Senator Patrick Moynihan stated that burning fossil fuels would cause "irreversible climate change," lead to a ten-foot rise in sea levels, and inflict "apocalyptic change."

In pre-hearing briefs the plaintiffs argued that the founding tenets of our democracy justified addressing climate change in the courts. The kids' lawyers pointed to the checks and balances among the legislative, executive, and judicial branches. Underscoring the venerable role the courts play in checking Congress and the president, the plaintiffs reached all the way back to 1789. "As James Madison stated when he presented the Bill of Rights to the Congress: 'If [these rights] are incorporated into the Constitution, independent tribunals of justice will consider themselves in a peculiar manner the guardians of those rights; they will be an impenetrable bulwark against every assumption of power in the Legislative or Executive; they will be naturally led to resist every encroachment upon rights expressly stipulated for in the Constitution by the declaration of rights.'"

Duffy and Sorenson further argued that this case didn't belong in the courtroom because these kids didn't belong in the courtroom. Duffy and Sorenson weren't accusing the kids of misbehavior or such. (Sure, a few of the youngest ones started fidgeting and doodling as the two-hour affair wore on, but they did so quietly.) Duffy and Sorenson meant the kids didn't belong in the legal sense. According to the defense, the kids lacked "standing." This legal concept, like many legal concepts, can get complicated, but basically it refers to a plaintiff's qualifications for bringing a lawsuit. It can be summed up in three questions: Have they suffered or will they imminently suffer a demonstrable injury? Can the defendant plausibly be blamed for the injury? Can a decision by the court likely provide a remedy to the injury? The government and the industry intervenors argued that the kids failed to qualify on all three counts.

Naturally, in the courtroom that day and in their prehearing briefs, the plaintiffs asserted that the kids did have standing. Without presuming to direct the government in detail, the plaintiffs outlined what they considered the necessary remedy. At some length they discussed what they see as the many reasons to lay a substantial portion of the blame for the injury at the feet of the government. But it is the first of the three standing questions—have the kids suffered or will they imminently suffer a demonstrable injury?—that received a great deal of attention from both the plaintiffs and the defendants and that resonates with the tenets of sustainability economics.

Let's consider the experience of one of the youth plaintiffs, Jayden Foytlin, who was twelve at the time of the initial filing of the lawsuit. Like the other kids, she mentions a variety of past, present, and future injuries she claims she had suffered, is suffering, and would suffer due in part to climate change. The harms the kids note range widely and include the aggravation of allergies, choking on smoke from wildfires, the erosion of nearby beaches, living next to a stinking expanse of rotting seaweed, an increased danger of shark attacks, a reduction in productivity on a farm, and having to move because of drought.

Due to the fact that Foytlin and her family live in the watery world of southern Louisiana, most of the problems she cites stem from the effect global warming is having on the nearby Gulf of Mexico: amplified hurricanes, sea level rise, and flooding from revved-up rainfall. Her family moved to their small town in Louisiana in 2005, and during the decade between then and the filing of the lawsuit, they got battered by three hurricanes and numerous tropical storms. In the document she submitted to the court she recounts how they lost power and water for a week in 2008 when Hurricane Gustav struck. In the 2015 complaint filed by the plaintiffs Foytlin also makes it clear that she worries about such impacts in the future. As it turns out, she was right to worry. In 2016, a flood engulfed part of her town, including her house.

In written declarations to the court, Foytlin and some of the other youth gave dramatic accounts of their tribulations, but dramatic stories don't ensure standing. The following passage from a pre-hearing document filed by the industry intervenors captures some of their main arguments against conferring standing on the kids. "The fundamental deficiency of these allegations is reflected in the fact that they are not in any way limited, or 'particularized,' as to these plaintiffs. The complaint asserts that the plaintiffs have standing to sue because they may in the future experience effects of climate change, and it identifies as those effects nearly every climatological, economic, epidemiological, meteorological, and political occurrence on the planet." The intervenors added, "Under the plaintiffs' view, anyone who may in the future suffer from any of those effects—in other words, anyone on the planet—could bring suit to force adoption of regulations that, in that person's view, are reasonably warranted to address those risks. This is plainly not a valid theory of standing. It is not enough for a plaintiff to allege that a defendant's conduct may generally contribute to a risk to 'society' or humanity in general."

In several ways the industry intervenors mislead or go overboard. For example, the plaintiffs do claim past, ongoing, and imminent injuries, not just the more distant future losses the intervenors mention. The intervenors also lightly mock the plaintiffs for claiming to have standing due to injury from "nearly every . . . occurrence on the planet," yet the science tells us that the long lists of effects the intervenors include are indeed happening and will get worse.

Still, the above flaws in their criticisms notwithstanding, the intervenors make some legitimate points. The courts seem built to decide specific cases, such as a nice, tidy inheritance dispute; you can fit the adversaries in a room, and you know they're fighting over the money and Mother's heirloom china. The kids' case is messy, sprawling. As the intervenors write, "It is the archetypal example of an 'abstract' and 'generalized grievance' that cannot support standing." Securing standing for the kids looked like an uphill battle.

On the other hand, picture Foytlin standing in sewage-laced floodwaters in her bedroom, as she did during the 2016 flood. Her injury was not "abstract" or "generalized." The impacts of climate change eventually work their way down to individual people. But the cause and effect are not as obvious as in a game of Clue in which Colonel Mustard kills someone in the conservatory with a revolver. That lack of clarity muddies court proceedings. It reminds me of the difficulties our market economy has in dealing with ecosystem services and other public goods. Note the parallels in some of the language used by the industry intervenors in their pre-hearing documents. For example, "Most of the adverse impacts alleged are to the environment or the public generally, rather than the plaintiffs personally." Or, citing some precedents, the intervenors write, "Those cases, like this one, involved claims by citizen-plaintiffs to prevent alleged waste or misuse of an asset held in trust for the public at large, and they were deemed non-justiciable because the claims addressed 'essentially a matter of public and not of individual concern.'" A few lines later the intervenors add, "The essential legal injury asserted here—that is, damages to a natural resource, the atmosphere, allegedly held in trust for the public—is shared equally by each and every other citizen." To my untrained ear the fact that damages extend to the whole world sounds like a compelling reason for a court to get involved, but the legal realm yearns for good old-fashioned individual suffering.

So how are these kids supposed to get equitable treatment in the courts for themselves and, by extension, for everyone else, including future generations? One path involves the claiming of constitutional rights, claims that the plaintiffs and defendants debated at great length. But the kids made an additional claim that might secure the broad intergenerational equity they seek, a claim discernable in the previous quote from the intervenors when they refer to assets held in trust for the public.

USUFRUCT

"Usufruct." Put an exclamation mark after this odd word and it sounds like fictional cussing you might hear in a sci fi movie, an invented but vaguely familiar-sounding expletive meant to convey that humans still curse when they hit their thumbs with hammers in a galaxy far, far away. But in the reality of earth in the early twenty-first century, "usufruct" is a venerable legal term that means the right to use another's property for a limited time without damaging or diminishing it.

"The earth belongs in usufruct to the living." That sentence comes from Thomas Jefferson, penned in 1789 in a letter to James Madison in which Jefferson expounds on the idea that the "fundamental principles" of every government rest on the concept of usufruct. Jefferson goes on to write, "The present generation of men . . . have the same rights over the soil on which they were produced, as the preceding generations had. They derive these rights not from their predecessors, but from nature."—that is, we the people, and therefore our elected government, have the responsibility to take care of the natural bounty that sustains us and pass it on intact to future generations. The idea of usufruct captures an essential element of the public trust doctrine, which in turn is an essential element of the kids' lawsuit.

I borrowed the above definition of "usufruct" and the Jefferson excerpts from *Nature's Trust*, the definitive nonscholarly book on public trust doctrine. Mary Wood, the author, teaches law at the University of Oregon and has been a major contributor to the thinking that informs the kids' lawsuit. And she has a vision for public trust.

Looking around at our deteriorating environment, emphatically including climate change, Wood thinks the U.S. government has permitted much of this destruction. Literally permitted it in many cases by issuing permits under existing environmental laws that allow unsustainable resource extraction and degradation. Instead of endlessly fighting in court over the details of such permits and statutes, she wants to go all in.

She advocates a transformational embrace of the public trust doctrine, the essence of which dates back to classical Rome. Echoing both Jefferson's thinking about usufruct and sustainability economics' perception of the environment as the foundation on which civilization builds, Wood sees this doctrine as enshrining a livable environment as a natural right that no generation can legally steal from another even if a legislature or executive permits the theft. As in *Juliana v. United States*, governments typically argue that legislative bodies, not the courts, should deal with broad environmental issues. Here's what Alfred Goodwin, a circuit judge for the U.S. Court of Appeals for the Ninth Circuit, writes about that notion: "As a coequal branch of government, the third branch must enforce the legislature's obligation to preserve the public trust." Pretty much everyone, including Wood, thinks it would be best if Congress and the executive branch took care of climate change, but, as Goodwin states, when the other two branches fail to do their duty, the courts must step in.

The role of the judiciary looms large in our current political climate, especially at the federal level. With the Trump administration and a GOP-controlled Senate waging war on environmental protection, the courts offer the best short-term hope for limiting the damage. (Elections offer the best long-term hope. For one thing, they might empower politicians who could restrain and eventually reverse the Trump/GOP push to appoint droves of hard-right judges to the federal bench.) However, the judiciary will not effectively blunt the environmental assault if they're overly deferential to the other two branches. Take the status of federal climate change policy. Customarily, the judicial branch defers to the climate expertise of the executive branch, particularly the EPA, whose primacy regarding global warming has been ratified by the Supreme Court. But now that Trump has appointed a former coal lobbyist to head the EPA, a strong public trust doctrine would require the courts to be more assertive in their oversight of the agency's climate policy.

Wood notes that scores of decisions by the U.S. Supreme Court and other American courts support the public trust doctrine, though she acknowledges that many decisions have gone against it, as well. An influential example of an affirming decision cropped up in the Pennsylvania Supreme Court in 2013. Overturning a state law promoting fracking, the chief justice quoted Pennsylvania's constitution: "The people have a right to clean air, pure water, and to the preservation of the natural, scenic, historic and esthetic values of the environment. Pennsylvania's public natural resources are the common property of all the people, including generations yet to come. As trustee of these resources, the Commonwealth shall conserve and maintain them for the benefit of all the people." The public trust sometimes extends to private property, too, when a particular use of that property would significantly harm vital public resources.

Wood likens the government's duty to that of a fiduciary whose job as trustee is to faithfully shepherd her client's assets—put the "trust" in "trustee." In this case the assets are nature's crucial resources and functions, and the clients are all of us. In her view, when government action damages those assets, it violates public trust. In addition, according to Wood, this fiduciary responsibility requires active protection of those natural assets, making it illegal for the government to do nothing or do less than enough to prevent "substantial impairment," to use the common legal term.

The fiduciary perspective evokes the SCC and the efforts to estimate the worth of ecosystem services. Both processes convey the value of natural assets, of public properties. Courts have always been quick to protect private property, likely influenced by an economy focused on market goods. Shoplift an iPhone and the law nails you. But courts can struggle with cases involving public goods because the pertinent laws lack clarity or may even support harm to public property, such as a stable climate. The kids maintain that the federal government behaves illegally when, for example, it leases so much public land and coastal waters for coal mining and oil drilling that it constitutes an excessive contribution to global warming. But existing statutes allow some such use of public lands, though environmental attorneys often sue over particular uses and sometimes prevail. It is the public trust doctrine that in one sweeping stroke, according to the plaintiffs, makes excessive mining and drilling illegal. Taking a page from sustainability economics, the kids implicitly insist that essential public goods, such as a stable atmosphere, are at least as valuable as iPhones, and courts should treat these natural assets accordingly.

Some observers concerned with the equity implications of the kids' lawsuit see a moral duty in public trust doctrine. Consider this excerpt from the "friend of the court" brief regarding the kids' lawsuit submitted by the Global Catholic Climate Movement (GCCM) and the Leadership Council of Women Religious (LCWR): "In raising the public trust doctrine, plaintiffs invoke the same moral imperative that

motivates the GCCM and LCWR. The public trust principle of law mirrors a sacred trust based on deep covenants
of obligation towards future generations and to all Creation.
Pope Francis described a sacred trust when he said, 'Creation
is not some possession that we can lord over for our own pleasure; nor even less, is it the property of only some people . . .
[C]reation is the marvelous gift that God has given us, so that
we will take care of it and harness it for the benefit of all.'" A
few lines later, the brief says, "The foundational U.S. Supreme
Court public trust cases hold that government has no authority to substantially impair or alienate resources crucial to the
public welfare. The Nation's public trust over these resources
remains an attribute of sovereignty that government cannot
shed. At a time when the climate crisis threatens the future
survival of civilization, the principle could hardly have a more
compelling application."

The government and industry intervenors kept their distance from the moral, fiduciary, environmental, and usufruct
arguments mustered by the kids in support of applying the
public trust doctrine to their case. Instead, the government's
only argument and industry's main argument against public
trust holds that precedent makes public trust strictly a matter for state governments, not the federal government. Possibly that's a legally accurate point, but whether accurate or
not, it feels like a technicality in the pejorative sense, one that
enables the defendants to dodge the heart of the debate.

In their closing arguments, Duffy and Sorenson presented straightforward summaries of their main legal points.
Olson likewise touched on her key ideas but with considerable passion, her belief plain to see in her delivery and words.
"Defendants are wrong that our complaint fails to allege constitutional and public trust violations for the harms caused
these young plaintiffs," said Olson. "Defendants in essence
ask this court to ignore the undisputed scientific evidence,

presented in our complaint and in opposing this motion, that the federal government has, and continues to, damage plaintiffs' personal security and other fundamental rights."

Shortly after Olson wrapped up, Coffin brought down the gavel and ended the hearing. Stirred up by some mix of the excitement of the proceedings and simple release from confinement, the youth plaintiffs were laughing and talking animatedly as they surged from the courtroom and out onto the courthouse steps for a post-hearing press conference. A couple hundred supporters enveloped them, cheering and applauding as several of the kids took turns at the mic. However, the celebration was premature, as Coffin had not yet rendered his decision.

But a month later the kids and their boosters had good reason to celebrate. Defying the odds, David beat Goliath. Coffin ruled against the motion to dismiss and recommended that the case move on toward trial. He found potential merit in the overall assertion that federal action and inaction regarding climate change constitutes inequitable treatment that favors the current generation over the kids' generation and the generations to follow. I won't drag you through the entire twenty-four-page decision, but here are some highlights that relate to issues we covered above.

Coffin declared that this dispute over climate change belonged in the courts and wasn't strictly confined to the legislative and executive branches. He writes, "The intractability of the debates before Congress and state legislatures and the alleged valuing of short term economic interest despite the cost to human life, necessitates a need for the courts to evaluate the constitutional parameters of the action or inaction taken by the government."

He found that the kids also belonged in the courts, rejecting the defendants' contention that the plaintiffs' alleged harms were too general and abstract to provide standing. He writes, "While the personal harms are a consequence of the alleged broader harms, noted above, that does not discount the concrete harms already suffered by individual plaintiffs or likely to be suffered by these plaintiffs in particular in the future."

As for the sweeping remedies sought by the kids, Coffin ruled them plausible and worthy of evaluation in a subsequent court proceeding. "Regulation by this country, in combination with regulation already being undertaken by other countries, may very well have sufficient impact to redress the alleged harms."

The judge determined that the public trust doctrine likely applies to the case. "The doctrine is deeply rooted in our nation's history and indeed predates it."

Perhaps most important, as it goes to the heart of *Juliana v. United States*, Coffin ruled that climate change may in fact violate the kids' constitutional rights. "If the allegations in the complaint are to be believed, the failure to regulate the emissions has resulted in a danger of constitutional proportions to the public health."

It's fitting that the last word goes to Kelsey Juliana, the lead youth plaintiff. Upon hearing the verdict, she responded that "this decision marks a tipping point on the scales of justice. Youth voices are uniting around the world to demand that Government uphold our constitutional rights and protect the planet for our and future generations' survivability. This will be the trial of the century that will determine if we have a right to a livable future, or if corporate power will continue to deny our rights for the sake of their own wealth."

But, of course, that is not the last word. As Juliana notes in her comment, the "trial of the century" is yet to come. Coffin's decision merely propelled *Juliana v. United States* to the next hurdle, a review of Coffin's decision by a fellow jurist in the same court, Judge Ann Aiken.

In November 2016, Aiken upheld Coffin's ruling. But "upheld" is much too bland a word for her resounding decision. In denying the government's attempt to get the case dismissed, Aiken supported the youths' constitutional right to a stable climate system and agreed that the public trust doctrine does indeed apply to their case.

Aiken also took the government to task for trying to steer the case away from the monumental issues it raises. In her decision she writes that the defendants "are correct that plaintiffs likely could not obtain the relief they seek through citizen suits brought under the Clean Air Act, the Clean Water Act, or other environmental laws. But that argument misses the point. This action is of a different order than the typical environmental case. It alleges that defendants' actions and inactions—whether or not they violate any specific statutory duty—have so profoundly damaged our home planet that they threaten plaintiffs' fundamental constitutional rights to life and liberty."

Finally, Aiken stakes out a strong role for the courts and reminds them that they are one of the three pillars that hold democracy aloft. She writes:

A deep resistance to change runs through the defendants' and intervenors' arguments for dismissal: they contend a decision recognizing plaintiffs' standing to sue, deeming the controversy justiciable, and recognizing a federal public trust and a fundamental right to a climate system capable of sustaining human life would be unprecedented, as though that alone requires its dismissal. This lawsuit may be groundbreaking, but that fact does not alter the legal standards governing the motions to dismiss. Indeed, the seriousness of plaintiffs' allegations underscores how vitally important it is for this Court to apply those standards carefully and correctly.

Federal courts too often have been cautious and overly deferential in the arena of environmental law, and the world has suffered for it.

Aiken then cites Judge Goodwin, whom we met a few pages ago. Goodwin writes: "The current state of affairs . . . reveals a wholesale failure of the legal system to protect humanity from the collapse of finite natural resources by the uncontrolled pursuit of short-term profits . . . [T]he modern judiciary has enfeebled itself to the point that law enforcement can rarely be accomplished by taking environmental predators to court."

You may recall another noteworthy event that occurred in November 2016: Donald Trump was elected president of the United States. When he and his administration assumed office, they became the defendants in *Juliana v. United States*. The Obama administration had filed the original motion to dismiss the lawsuit and avoid a trial, but when Trump and his many fossil-fuel-connected appointees took over the defense, they shifted the evasive maneuvers into high gear. Writs of mandamus, an interlocutory appeal, a motion for a protective order and a stay of all discovery, a motion for judgment on the pleadings, a motion for summary judgment—the administration deployed its entire arsenal of avoidance weaponry. The battles over all these writs and motions ranged back and forth from the District Court for Oregon to the Ninth Circuit Court of Appeals and back to the District Court and back to the Ninth Circuit and to the U.S. Supreme Court and back to the District Court and back to the Ninth Circuit . . . well, you get the idea.

Tired of waiting, Levi Draheim issued a statement bemoaning all the delays. "We've known about the dangers of climate change for over fifty years and can't wait two more years to go to trial to stop it. I filed the lawsuit when I was eight years old and now I'm eleven, and now we might have to wait until I'm fourteen. I'm seeing the effects of climate change happen all around me in Florida; we don't have two more years to lose."

In the course of this judicial wild ride, *Juliana v. United States* has become a national and global cause célèbre. At the first hearing, one local TV crew showed up to shoot a few minutes of footage. Now the media swarms any event related to the lawsuit. In the months before the time of this writing, coverage of the youth showed up on *60 Minutes*, CNN, CBS News, NBC News, and ABC News. Articles appeared in *The New Yorker*, *The Washington Post*, *Forbes*, *The Wall Street Journal*, *The New York Times*, *Time*, and *Newsweek*. Coverage also spread abroad to outlets such as *Le Monde*, *The Guardian*, Reuters, *News Ghana*, *The Australian*, and *The Irish Times*.

A flurry of activity has accompanied the media furor, including the filing of fifteen more amicus briefs in support of the youth lawsuit. Three U.S. senators and four members of the House filed a friend of the court brief. So did eighty-two law professors, ten businesses and business organizations, seventy-eight doctors, and fourteen health organizations, including the American Lung Association, the American Pediatric Society, and the American Heart Association. In 2019, Zero Hour, a youth-led climate organization, set up a website to gather signatures to support the "Young People's Brief." By the time the brief was submitted to the Ninth Circuit, more than 36,000 young people had signed on.

This outpouring of support by no means guarantees victory for the youth. In fact, they suffered a serious setback on January 17, 2020, when a three-judge panel of the Ninth Circuit ruled by a two-to-one vote to dismiss the case. This was not the end of the line—the plaintiffs plan to appeal—but most observers think the youth will lose that appeal and the case will never proceed to trial.

Even if, as expected, the January 17, 2020, ruling of the Ninth Circuit turns out to have been the beginning of the end, the text of the decision reveals how much the kids have accomplished and how the legal and political winds regarding climate change are shifting. Consider the opening words of the decision: "In the mid-1960s, a popular song warned that we were 'on the eve of destruction.' The plaintiffs in this case have presented compelling evidence that climate change has brought that eve nearer. A substantial evidentiary record documents that the federal government has long promoted fossil fuel use despite knowing that it can cause catastrophic climate change, and that failure to change existing policy may hasten an environmental apocalypse."

Here's the kicker: that quote comes from Judge Andrew Hurwitz, writing for the majority—the majority being he and Judge Mary Murguia, the two judges who ruled that the case should be dismissed. They repeatedly make clear that they think the youth presented convincing arguments. For example, the decision declares, "The plaintiffs have made a compelling case that action is needed; it will be increasingly difficult in light of that record for the political branches to deny that climate change is occurring, that the government has had a role in causing it, and that our elected officials have a moral responsibility to seek solutions."

The reason Hurwitz and Murguia called for dismissal, despite their evident sympathy for the youth's cause, is their belief that the plaintiffs lacked one element of standing; the two judges felt the courts could not provide a remedy and that climate change should be addressed by Congress and the executive branch. As the decision notes, "We reluctantly conclude, however, that the plaintiffs' case must be made to the political branches or to the electorate at large, the latter of which can change the composition of the political branches through the ballot box. That the other branches may have abdicated their responsibility to remediate the problem does not confer on [certain federal courts], no matter how well-intentioned, the ability to step into their shoes."

The third person on the Ninth Circuit panel, Judge Josephine Staton, passionately dissented from the decision to dismiss. She begins her opinion by writing, "In these proceedings, the government accepts as fact that the United States has reached a tipping point crying out for a concerted response—yet presses ahead toward calamity. It is as if an asteroid were barreling toward Earth and the government decided to shut down our only defenses. Seeking to quash this suit, the government bluntly insists that it has the absolute and unreviewable power to destroy the Nation."

Though Congress and the executive branch are best suited to handle a matter like climate change, they have known about the dangers since at least 1965, as the majority opinion notes, and have done little to protect the country from that approaching asteroid while doing much that has accelerated it. Staton understands that overseeing the development and implementation of a remedy for a massive problem like climate change would not be easy for the court, but she asserts that it can and must be done. She points to the landmark U.S. Supreme Court case *Brown v. Board of Education,* which ordered that America's public schools be desegregated. That was a vast, ambitious undertaking, yet the court waded in because constitutional rights were being violated and Congress and the executive branch weren't stepping up.

You may recall that in this book's introduction I asserted that the environment is the bedrock on which all other elements of civilization rest. In her dissent, Staton views *Juliana* through a similar lens, declaring that climate change threatens the perpetuation of America as an orderly nation and therefore threatens our constitutional rights, which rely upon the continuing health of our society. "Much like the right to vote, the perpetuity of the Republic occupies a central role in our constitutional structure as a 'guardian of all other rights.' Civil liberties, as guaranteed by the Constitution, imply the existence of an organized society." And if the climate grows too unstable, society might descend into chaos, and our liberties would melt away like so much overheated polar ice.

In her conclusion, Staton returns to *Brown v. Board of Education* to note that in one crucial sense even such imperative social issues as school desegregation are less urgent than climate change. In her closing words you'll hear echoes of our earlier discussion of the implacable nature of biophysical reality; just as markets struggle to grasp its significance, so apparently do our courts. Staton writes, "Were we addressing a matter of social injustice, one might sincerely lament any delay, but take solace that 'the arc of the moral universe is long, but it bends towards justice.' The denial of an individual, constitutional right—though grievous and harmful—can be corrected in the future, even if it takes 91 years [the time between the Emancipation Proclamation and the *Brown* decision]. And that possibility provides hope for future generations. Where is the hope in today's decision? Plaintiffs' claims are based on science, specifically, an impending point of no return. If plaintiffs' fears, backed by the government's own studies, prove true, history will not judge us kindly. When the seas envelop our coastal cities, fires and droughts haunt our interiors, and storms ravage everything between, those remaining will ask: Why did so many do so little?"

6

TOMORROW EVENTUALLY
BECOMES TODAY

With all the legal jargon and constitutional concepts flying around the courtroom during the initial hearing of *Juliana v. United States*, few people noticed an economics term that momentarily surfaced that March morning. While questioning Sean Duffy, the government's attorney, Judge Coffin wondered if the government was dumping the social costs of climate change onto the kids and future generations and giving the social benefits to present generations of adults. He asked, "Are you robbing Peter to pay Paul?" In the course of this discussion he mentioned something called the "discount rate."

Judge Coffin did not elaborate on the discount rate, but we're about to because it gets to the heart of intergenerational equity and is probably the single most decisive factor in calculating the social cost of carbon.

The discount rate adds a new dimension—a literal dimension, time—to the already roiling SCC debate. The key point of contention is how much society values the present compared with the future. This tug-of-war between the present and the future involves complex decisions about the equitable distribution of resources among generations and profound disputes about the role of ethics in economics.

Basic finances underpin a traditional view of discounting. If you invest $100 at a 5 percent annual interest rate, in one year you'll have $105. If you let that investment ride for ten years, you'll have $163 thanks to the magic of compound interest. Well, discounting works like compound interest in reverse. For example, at a 5 percent discount rate, $100 due ten years in the future is worth about $60 today. By flipping this "time value of money," economists can calculate future benefits and costs into today's dollars, which enables them to assign a value today to a benefit or cost that will occur in the future. This assigned value can make or break a project like a community solar array. A high discount rate would shrink the estimates of its future benefits and perhaps convince decision-makers that the array isn't worth the upfront expense.

Other less mathematical factors also can influence the discount rate. For one thing, we humans generally want things sooner rather than later. Much of this desire for sooner stems from simple eagerness—I want that car or phone or sofa or marshmallow ASAP. I added marshmallow to that list in honor of the famous marshmallow experiment, in which researchers test the patience of children around the ages of four or five. Typically, a researcher will seat a kid alone at a table in an otherwise empty room and set a single marshmallow in front of the salivating youngster. The researcher then says she has to leave the room briefly but promises the kid two marshmallows instead of just the one if the kid can refrain from eating the single marshmallow while the researcher is gone. Then the researcher leaves and the kid . . . well, you should watch one of the online videos that show the ensuing torment.

The kids stare longingly at the tempting treat. They sniff it, poke it, caress it, press it to their lips. They squirm and bounce and grimace and writhe. Eventually most of them succumb, maybe starting with a nibble or a furtive bite before wolfing down the whole marshmallow. But some manage to resist the entire time. One video chronicles the tribulations of a boy whose expressions of suffering surpass those of Al Pacino at his overacting best, but the boy actually went the distance, even though at one point he held the marshmallow inside his open mouth. When the researcher returned and gave the boy his reward, the kid snatched the two marshmallows and stuffed them into his mouth with both hands. Of course, some kids gobble the marshmallow as soon as the researcher leaves the room. One red-haired girl started eating her marshmallow while the departing researcher was still saying good-bye. Clearly, we humans like to have our cake and eat it soon.

Putting a number on this tendency to value the here and now over the there and then is shaky, especially in the context of issues like climate change. In such situations economists, particularly sustainability economists, often use the term "social discount rate" to distinguish it from discounting done by individuals, whose concerns reach no further than their own lifetimes and bank accounts. But societies live indefinitely, and sustainability economists think that when dealing with social issues, we should use discount rates that consider future generations and society's general welfare.

Another factor that favors the present stems from the fact that the economy has grown enormously and fairly steadily since the advent of the Industrial Revolution, leading many economists to assume that on average future people will always be richer than present people—vastly richer in most estimations. And if you become richer, then each dollar matters less to you. Therefore, goes the reasoning, a dollar now is worth more to you than a dollar in the future is worth to the average denizen fifty or one hundred years from now, who will be wallowing in wealth. In this case you should deeply discount the future. As with our desire to have things ASAP, putting a number on this belief in a richer future is tricky, but most economists do so, and it is vital in shaping the discount rate.

Above I referred to the magic of compound interest, but the example I used—investing over a ten-year period—does not do justice to that magic. To appreciate the explosive force of compound interest—and discounting—we need to look further into the future. Let's start with $100 at 5 percent for one hundred years. That comes to about $13,000. But this book deals with the social cost of carbon, and carbon dioxide lingers in the atmosphere for hundreds and even thousands of years, so let's ponder the effect of compounding over a longer stretch of time, say five hundred years. And let's shift our units from dollars to deaths. Moral philosopher Derek Parfit satirically notes that "at a discount rate of 5 percent, one death next year counts for more than a billion deaths in 500 years. On this view, catastrophes in the further future can now be regarded as morally trivial."

Finally, to fully convey the awesome exponential power of discounting, let's consider what happens over the course of a couple of millennia. Here's an example from a *Science News* article by Julia Rehmeyer: "At a 5 percent annual interest rate, a penny that belonged to Julius Caesar would have expanded to [a sum of money greater] than the entire world economic output over the last 2,000 years multiplied by the number of stars in the sky. So the brutal arithmetic of discounting (at a 5 percent social discount rate) would shrink any imaginable catastrophe today to far less than a penny in Caesar's time, and an economist would have therefore recommended that Caesar not spend even so tiny an amount to avoid it."

Here's what the "brutal arithmetic" of discounting means in the real world: a high discount rate can dissuade society from spending, say, a few trillion dollars on climate action today even if that action would save tens of trillions in social costs over the next several decades or hundreds of trillions over the next couple of centuries. Seduced by compound interest, conventional financial thinkers frown upon spending money in the present to reduce emissions for the sake of the future. Instead, this pinched logic concludes that we would be better off if we invested that money and kept growing our wealth rather than spending it on climate action. As we explore discounting, keep in mind two questions regarding the above phrase "we would be better off." Who is "we"? What does "better off" mean?

"There's very little agreement on what the discount rate should be and very little agreement on how we should come up with it, but it's probably the single most important variable in determining the social cost of carbon," said Josh Farley, the University of Vermont economist and coauthor of *Ecological Economics*. "If you have a discount rate of 6 percent, that means every twelve years, the value of something falls by half." He added, "This is the main reason most economists are very complacent about climate change. They say that the costs of addressing it now exceed the benefits we'll get in the future because we discount those benefits."

Discount rates incited fierce debate within the interagency working group in 2009, when the group developed its initial SCC. (Note that the working group, not the IAM developers, chose the rates.) "We were fighting and fighting about the discount rates," said Dina Kruger, the director of the EPA's Climate Change Division in 2009 and an IWG member. "DOE [Department of Energy] and EPA were fighting for a discount rate of 2 percent, and [those of us thinking about it in the context of intergenerational discounting] were saying it should probably be 1 percent or 1.5 percent." On the other hand, the more market-oriented members were looking at rates from 3 percent to 7 percent.

The controversy partly stems from the clash between two

views of the role of equity in climate economics. One view holds that the market should set discount rates. Some proponents of this perspective propose a 7 percent rate for the SCC because many analysts figure that's the rough historical average for the return on investment in the U.S. stock market. According to this camp, investors are essentially demanding a 7 percent return before investing their money and thereby surrendering the opportunity to use it for other purposes, which reveals their price for waiting to eat that marshmallow.

Others who look to market behavior for answers think the discount rate used for the SCC should be based on consumer behavior, not the behavior of capital, given that the SCC attempts to measure the loss of consumption due to climate change. Armed with a variety of calculations and assumptions, proponents of this consumer-driven framework typically select something in the vicinity of a 3 percent rate. Whichever of the two market approaches they support, people in either camp see discount rates as something supply and demand provides, not a judgment call involving ethics.

The other view of equity's role, more in line with sustainability economics, holds that ethics should play a major role in discounting. Advocates of this perspective believe the present generation has an equal but not a greater right to resources than future generations. Many see a place for the consumption-based market approach and minimal discounting—though some think the rate should be zero—but they consider the market angle subordinate and instead see providing equal resources to future generations as an overriding moral imperative. "At its heart it's a value statement about the weight you're going to put on the welfare of current generations versus future generations," said Kristen Sheeran, the climate change policy adviser to Oregon's governor.

In the end, the IWG officially tilted toward the market-based, consumption-oriented approach. They acknowledged the complexities of discounting by presenting a range of rates, including 2.5 percent, 3 percent, and 5 percent, with 3 percent as their central rate—the one that would most influence policy and legal proceedings. They got an earful of dissenting views from outsiders as well as from within their own ranks, but the leaders of the group felt that the consumer-behavior approach was the "most defensible and transparent," as they put it in an IWG document. A 3 percent rate certainly justifies working much harder and sooner to curb global warming than does a 7 percent rate, which would justify spending little more than one of Caesar's pennies to forestall climate change. (Few people propose 7 percent, but as we'll see later in this chapter, those few wield a lot of power.) But many observers, especially those with an eye for equity, think that 3 percent—and a range from 2.5 percent to 5 percent—is still too high. They think the market provides dubious guidance and should take a backseat to ethics.

Market ideals themselves undermine the market approach to discounting. The basic principle of supply and demand holds that consumers and producers create the best economy through their choices in the marketplace, but the consumers and producers of future generations aren't here to make choices. We residents of the present are making all the choices for them. As the marshmallow test demonstrated, humans tend to favor the present over the future, but if we in the present eat all the marshmallows, then future folks won't *get* to choose whether to eat the one marshmallow immediately or to wait and get two. Because they're not with us in the here and now, they will be unable to vote or lobby or otherwise take part in decisions that will affect them deeply, which seems un-American, like taxation without representation.

But let's set aside future generations' absence from today's markets and examine the reason economists most often cite for using a higher discount rate for the SCC: they think endless economic growth will make future generations much wealthier than the present populace and therefore better able to afford to address climate change. Given this sanguine outlook, some economists think the market approach and a higher rate do, in fact, serve equity. After all, they ask, what's fair about asking the poorer people of the present to preserve resources for the richer people of the future?

This overconfidence in our financial future results in a view of discounting that can warp the response to climate change. Remember Martin Weitzman from the chapter on fat tails, he of the dismal theorem? You may recall that he once took a climate IAM and extrapolated from it to produce a scenario in which an outlandish global temperature rise resulted in a huge GDP loss. Weitzman used the IAM's assumptions, including its moderate discount rate, to calculate what would happen if the earth heated up 10°C over two centuries. The IAM estimated this would cause a 19 percent drop in GDP. (As noted in the "Fat Tails" chapter, many economists and pretty much every climate scientist think that a rise of 10°C would ravage the planet and produce a GDP loss far greater than 19 percent. However, for our purposes here, we'll go with 19 percent, which would still constitute a massive economic blow.)

In his scenario, Weitzman also stipulated that during those two hundred years the world's economy would chug along at a moderate annual growth rate of 2 percent. Via the rosy math of discounting, that growth would render a 19 percent hit inconsequential. Weitzman's calculations reveal the Pollyanna view that even the drag on the economy caused by such extreme global warming would only slightly slow GDP expansion; instead of growing fifty-five times larger in those two hundred years, the economy would grow a mere forty-four times larger.

Not exactly hard times. Maybe some of our descendants would have to cut back a little on their caviar intake, but the specter of such future deprivation isn't likely to motivate those of us living in the present to spend much of our relatively puny GDP on reducing greenhouse gas emissions to benefit filthy-rich future generations. This vision of an inexorable march to wealth is an illusion that we'll examine in detail in a later chapter. In the meantime, we should not let this fantasy fool us into believing that even with extreme warming the magnitude of damages will be brushed aside by the dazzling wealth that inevitably awaits our descendants.

What if it were true that money not spent now on reducing greenhouse gases would be invested and balloon into big piles of money that could then be spent by enriched future generations to take care of climate change? Problems would still arise.

Yes, those piles of money *could* be used to address climate change, but would they be? If we don't shift to a sustainable economy, it seems unlikely that society would in fact spend those piles of money on addressing global warming. More likely, the market would continue steering most resources toward consumer goods rather than a stable climate and other public goods.

The influence of market failures on discounting goes deeper, too. A chunk of those ballyhooed returns of 7 percent relies on the exploitation of externalities. Those fat returns would undoubtedly slim down if businesses had to account for social costs. This means that in order for many investments to keep earning such high interest, they would have to continue exploiting externalities. Which means that past returns based on externality-riddled investments do not provide an accurate or ethical yardstick for determining discount rates.

COMPOUND INTEREST FELLS REDWOODS

The intertwining of discounting, investments, natural resources, and social costs brings to mind a house tour I once took in McKinleyville, a sparsely populated backwater on the far-north California coast. I got to talking with the owner of a six-thousand-square-foot mansion, and he invited me in to have a look. At every turn this ostentatious custom house boasted notable features such as the high-ceilinged, tennis-court-sized great room, which evoked a medieval lord's grand dining hall. But even more notable than the great room was the fact that this manor house had been built from the wood of a single tree, an old-growth redwood.

As that mansion suggests, old-growth redwoods are extraordinarily valuable. Most obviously, they're huge, often taller than two hundred feet (the tallest recorded was 379 feet) and measuring maybe ten to fifteen feet in diameter. Builders and homeowners also covet redwood for its resistance to water and rot. Old-growth redwoods come by that hyphenated adjective honestly; an average old-growth redwood alive today sprouted in the Middle Ages, and one of the elders might have started its skyward journey before Jesus walked the earth. Growing slowly over the ages, old-growth redwoods develop an exceptionally tight grain, which helps make them two or three times more valuable per board foot than younger redwoods.

Given their value, it wasn't long before lumber companies descended on the coastal redwoods, which grow in a narrow band from Central California to just north of the Oregon border. By the 1980s, loggers had clear-cut nearly all of these grand forests. Only a few pockets of old growth survived, most of them in the area that includes McKinleyville.

As the trees disappeared, so did nearly all the logging outfits, but one remained just a few miles south of McKinleyville. Founded in 1863 and run by the Murphy family for most of its existence, the Pacific Lumber Company had long prospered by treating its employees and its forests uncommonly well. Instead of removing mature trees faster than younger trees could replace them, Pacific Lumber cut at a sustainable pace. And instead of scything trees like a crop of corn, the company practiced selective logging, taking only a portion of a stand and leaving ample trees to maintain considerable forest function and to form the foundation for the new forest. They paid attention to other elements of the forests' ecosystems, too, such as taking care not to log slopes that would send silt down into salmon habitat.

But the allure of compound interest ruined this idyll. The 1980s were the era of the corporate raider, and one of the most notorious of the breed, Charles Hurwitz, engineered a hostile takeover of Pacific Lumber in 1986. Immediately his company, Maxxam Inc., revved up the rate of logging way beyond sustainable levels. And why not? He could invest his profits in ventures that earned better returns than a do-gooder lumber company earned puttering along at a pace that sustained its workers and the environment. All those ancient redwoods—Pacific Lumber owned the largest private expanse of old-growth coastal redwoods in the world—were worth more to him sheared and turned into investments than left standing.

This cold logic appalls many observers. Even people who have no specific knowledge of the nonmarket values being lost as the redwoods fall can see that the future is being sacrificed to the present. The people who do know about the value of ecosystem services have a more specific understanding of the extent of the loss. They know that not only are people losing the services for the following year but for every year for a long, long time, perhaps forever. Calculating a value for ecosystem services is hard enough, but when we are dealing with services that keep delivering indefinitely, it becomes impossible to precisely estimate their future value because it's essentially infinite. This suggests that we shouldn't discount the value of resources such as the public goods provided by redwood forests using the same financial tools we use to discount market goods.

When we are speaking of a stable climate or some other public good, a discount rate based on financial investments implies that money can substitute for lost ecosystem services or other natural capital. I dig deep into substitutability in a later chapter, but in the context of discounting and equity, I'll mention two things here. First, a rapidly compounding bank account can't restore the Greenland ice sheet if it melts. Second, how many of you think Hurwitz's returns from investing the profits he made by scalping the redwoods were used to compensate people for their loss of flood control, salmon habitat, and other forest services? Yeah, me, neither. Even if money could always substitute for nature, the notion that society gets richer when an individual gets richer, regardless of the use and distribution of those riches, violates basic principles of equity.

One more reason not to use the market as the primary method for choosing a discount rate: climate change itself. The impacts of global warming call into question the orthodox view of ever-growing future riches, which provides the main justification for adopting a high rate. This is another topic I'll explore in depth later, so for now I'll just point out that abundant studies predict that unaddressed climate change would ravage the economy and make the future poorer, perhaps much poorer, than the present.

The passionate debates within the IWG about the discount rate came as no surprise. Just three years earlier, in 2006, a much bigger discounting firestorm erupted after the publication of the Stern Review, the influential UK government report on the economic impacts of climate change that I mentioned earlier. Stern and company used a 1.4 percent discount rate, far lower than the typical 3 to 7 percent range. Largely due to that low rate, Stern came up with a central estimate of $85 a ton for the SCC, a figure many times higher than most of its contemporaries. If used for climate policy, such a high number would send the message that society should spend much more money on reducing emissions sooner rather than later.

Predictably, the fossil fuel industry pummeled the review, but they were not alone. "Stern was absolutely pilloried by the mainstream economic community," said Karl Hausker, a senior fellow in the World Resources Institute's Climate Program and an authority on the SCC. "They ripped him up and down, left and right. He dared to use a low discount rate, outside of the mainstream economic opinion. They ripped him about that in particular." The fervor of some of the criticism from Stern's fellow economists suggests that this was more than just an academic disagreement over discounting. His harshest critics seemed especially disturbed by his injection of an ethical perspective into what they like to think of as a purely empirical exercise about the workings of markets.

Here's an example of the kinds of statements in the Stern Review that agitated so many mainstream economists: "In drawing up a social welfare function, we have to make explicit value judgements about the distribution of consumption across individuals—how much difference should it make, for example, if a given loss of consumption opportunities affects a rich person rather than a poor person, or someone today rather than in a hundred years' time?" And a few paragraphs later, "We treat the welfare of future generations on a par with our own. It is, of course, possible that people actually do place less value on the welfare of future generations, simply on the grounds that they are more distant in time. But it is hard to see any ethical justification for this."

In their ecological economics textbook, Herman Daly and Josh Farley point out that higher discount rates based on investment returns seem better suited to individuals than to society. An individual may reasonably base his balancing of present versus future desires on his impatience to eat that marshmallow in front of him. Society, however, is immortal and should take the longer view and choose a discount rate "that reflects society's collective ethical judgment, as opposed to an individualistic judgment, such as the market rate of interest." Daly and Farley conclude their discounting discussion by writing, "Justice replaces efficiency as the relevant criterion for policy when time periods become intergenerational."

Stern has noted that many supposedly empirical examinations of climate economics include implicit value judgments, just not value judgments that favor intergenerational equity. "These discount rates are central to any discussion of our hand in the fate of future generations," writes Stern. "Most current models of climate-change impacts make two flawed assumptions: that people will be much wealthier in the future and that lives in the future are less important than lives now." He adds, "The former assumption ignores the great risks of severe damage and disruption to livelihoods from climate change. The latter assumption is 'discrimination by date of birth.' It is a value judgement that is rarely scrutinized, difficult to defend and in conflict with most moral codes."

Given the widespread criticism of Stern's discount rate and the resistance to lower rates that prevailed during the IWG's 2009 proceedings, it appeared that future generations were in for a rough ride, but in some places a groundswell of support for lower rates and higher ethics was building. For example, in 2012, Laurie Johnson and Chris Hope published an influential paper that took the IWG's discount rate to task. (At the time Johnson was an analyst for the Natural Resources Defense Council, and Hope, you may recall, is the creator of PAGE, one of the Big Three IAMs.) Based on both market and equity concerns, they figured the SCC should use much lower discount rates. Their use of lower discount rates led them to estimate that the SCC should be 2.6 to more than 12 times higher than the IWG's initial number of $21 per ton of CO_2. That would produce an SCC as high as $266, which would hugely boost the value of early and rapid reductions in greenhouse gas emissions.

Such subversive ideas emanated from other, even more high-profile sources, too, including the Intergovernmental Panel on Climate Change. Because it includes almost every nation in the world and must accommodate an immense range of viewpoints, the IPCC tends to flow with the mainstream. Yet in its 2014 report the IPCC recommended a healthy dose of ethics in its discounting decisions. The IPCC also endorsed the use of a concept that had been gathering steam in climate economics circles: a declining discount rate. Remember the absurd examples I mentioned earlier, such as the one in which a 5 percent discount rate would make it inefficient for Julius Caesar to have spent one penny to save us people of two thousand years later an essentially infinite amount of costs? Declining rates restore sanity to the exponential madness that occurs over long time periods by starting at, say, 3 percent and then ratcheting down over time. Some countries, such as the United Kingdom and France, use declining rates.

So the momentum began to swing toward lower and declining discount rates. But then Donald Trump became president and the seven-percenters assumed power.

THE 7 PERCENT SOLUTION

On March 28, 2017, President Trump signed Executive Order 13783 to "promote energy independence and economic growth." It mainly consists of a hit list of rules, guidance documents, memoranda, and other directives related to fighting climate change, many of Obama vintage, that Trump has commanded executive agencies to revoke or review. And context makes it clear that "review" means find a way to nix or neuter the items in question.

Partway down the list the order calls for a "review" of the SCC, effectively rescinding the current number. Eight directives related to the SCC follow. First, the executive order disbands the interagency working group. No longer will it be around to update and improve the SCC. Next comes a string of half a dozen key SCC documents "that shall be withdrawn as no longer representative of governmental policy." The erasure of these IWG materials means the executive branch has abandoned the scientific and economic analysis that underpinned the SCC. Finally, the last of the eight SCC directives includes instructions to use "appropriate" discount rates. The directive defines "appropriate" as rates that adhere to the guidelines found in the Office of Management and Budget (OMB) Circular A-4, published in 2003 by the George W. Bush administration. The Trump administration says that it will eventually develop a new SCC, but in the indefinite meantime, they refer all things SCC to Circular A-4, even though it's merely a general guide to regulatory analysis.

Circular A-4 instructs executive agencies and departments to conduct most analyses using both a 7 percent discount rate and a 3 percent rate, but the Trump administration clearly has a soft spot in its heart for 7 percent. For years fossil fuel

industry groups and like-minded think tanks have been lobbying for 7 percent. David Kreutzer, a senior research fellow
at the Heritage Foundation and prominent champion of the
7 percent solution, wrote in 2016, "To be done properly, the
discount rate should reflect the best rate of return that could
reasonably be expected in capital markets. Over the past two
centuries, the stock market in the U.S. has generated a return
of more than 7 percent . . . Therefore, the 7 percent discount
rate stipulated by the Office of Management and Budget
for benefit-cost analysis seems very appropriate for use in
analysis of climate policies." Kreutzer also served on Trump's
EPA transition team and as one of the architects of Executive Order 13783. Benjamin Zycher, a visiting scholar at the
American Enterprise Institute, likewise embraces 7 percent.
In 2016 he wrote, "The Obama IWG calculation of the SCC
is fundamentally dishonest for several reasons, only one of
which is the failure to use a 7 percent discount rate, despite an
OMB requirement to do so as outlined in Circular A-4. The
obvious reason for that refusal is the fact that the SCC falls to
zero or a negative number . . . at a 7 percent discount rate."

Suddenly the arguments about the discount rate among
members of the IWG or the criticisms by Johnson and Hope
seem like minor disagreements among generally like-minded
colleagues. Sure, there's a meaningful difference between 2
percent and 3 percent, but 7 percent? That would render the
SCC meaningless as a tool to curb emissions.

À la Trump's executive order, Kreutzer and Zycher cite the
OMB and Circular A-4 as the definitive source for discounting decisions. Note their exact words: Kreutzer writes that the
OMB "stipulated" the use of 7 percent, and Zycher writes
that using 7 percent is an OMB "requirement." Neither of
those two words is correct. Though George W. Bush famously
shunned nuance, his OMB's Circular A-4 actually contains a
fair bit of the stuff. Yes, the circular favors including 7 per-

cent (and 3 percent) in most situations, but not all; a 7 percent discount rate is not an across-the-board stipulation or requirement. In a passage on intergenerational discounting, the circular states, "Special ethical considerations arise when comparing benefits and costs across generations. Although most people demonstrate time preference in their own consumption behavior, it may not be appropriate for society to demonstrate a similar preference when deciding between the well-being of current and future generations. Future citizens who are affected by such choices cannot take part in making them, and today's society must act with some consideration of their interest." Note the word "must" in the previous sentence. Now *that* sounds something like a stipulation or a requirement.

So why such a push by fossil fuel interests and kin for 7 percent? No mystery. Zycher said it quite clearly: "The SCC falls to zero or a negative number . . . at a 7 percent discount rate." And there it is, the 7 percent solution that would help greenhouse gas emitters dodge the many climate change regulations that involve cost-benefit analysis.

Richard Revesz, a professor at New York University School of Law, director of the Institute for Policy Integrity, and an authority on cost-benefit analysis and environmental regulation, thinks the 7 percent discount rate is a key tool in the Trump administration's campaign to undermine the SCC. Revesz also noted that Trump's executive order does away with the previous SCC but does not make clear what, if anything, will replace it, apart from partial help from Circular A-4. "[The executive order] gives no guidance," he said. "Agencies are going to be all over the map in how they respond. The result is probably going to be confusion and inconsistency, which are likely to end up in judicial challenges to the decisions." Given that several courts upheld Obama's SCC and the process that produced it, Revesz thinks legal challenges to

Trump's haphazard treatment of the SCC will likely succeed. He said that the Institute for Policy Integrity is watching carefully. If agencies don't value the SCC appropriately or don't follow proper procedures, he said, "We will bring challenges in every single case. We're tracking every proceeding." For example, he thinks they could take the 7 percent solution to court and probably win.

The threat of legal challenges may be the main reason the Trump administration hasn't simply eliminated the SCC altogether. Since the *Center for Biological Diversity v. National Highway Traffic Safety Administration* case, several judicial opinions have affirmed the Ninth Circuit's seminal decision that the SCC can't be zero. They have also affirmed the legitimacy and credibility of the IWG's SCC. Knowing that the courts will most likely continue requiring regulations to account for the costs and benefits of climate change, the Trump administration has eschewed a frontal attack in favor of subverting the SCC by altering the ways in which it is calculated. The discount rate is one of the administration's two principal monkey-wrenching tools. The other fits perfectly with one of Trump's favorite themes, catchphrase and all.

AMERICA FIRST

Global warming is, well, global. We don't say "national warming." Whether emitted in Poland or Brazil, rising greenhouse gases end up in the one and only atmosphere we all share. This shared fate provides the basic ethical rationale for using a global social cost of carbon. A global SCC estimates the impact of a ton of carbon dioxide on everyone on the planet.

The interagency working group used a global SCC, as do most nations that employ a social cost of carbon. But the Trump administration has rejected the global perspective in favor of a domestic SCC. Trump made his underlying rationale clear in a speech about withdrawing the United States from the Paris climate agreement: "I was elected to represent the citizens of Pittsburgh, not Paris." Never mind that the residents of Pittsburgh will likely suffer just as much as Parisians from Trump's climate policies. Assuming the use of a domestic SCC survives legal challenges and is implemented, it means the federal government will only account for the impact that a ton of CO_2 has within the boundaries of the United States. Like a high discount rate, a domestic estimate drastically reduces the SCC. In the case of a proposed climate regulation that hinges on cost-benefit analysis, either a high discount rate or a domestic SCC can make its adoption unlikely. In combination, they make adoption nearly impossible.

Take the Clean Power Plan (CPP). Arguably the centerpiece of Obama's climate agenda, the CPP aimed to greatly reduce greenhouse gas emissions from power plants. Using the then-prevailing SCC, with its global scope and 3 percent central discount rate, the EPA estimated that by 2030 the CPP would return annual net benefits between $26 billion and $45 billion—far more than the cost to industry of abiding by the plan. But then Trump moved into 1600 Pennsylvania Avenue and appointed a staunch climate skeptic and fossil fuel booster, Scott Pruitt, to head the EPA. Before he resigned in scandalous ignominy a year and a half later, Pruitt devised a scheme to scuttle the CPP by, in part, proposing a new approach using a domestic SCC and discount rates of 7 percent and 3 percent. The IWG's updated SCC had climbed to about $40 just before Trump's executive order vaporized the working group. And remember, many experts considered $40 much too low. But by applying a domestic SCC and a 7 percent discount rate, Pruitt's recalculation sank the SCC to an impotent $1 ($6 at the 3 percent rate). As of this writing, the whole matter is up in the air; the Trump administration has issued a fossil-fuel-friendly plan intended to replace the CPP, but it faces tough courtroom battles. Predictably, the Trump plan features a domestic SCC and a 7 percent discount rate.

Pruitt is not the first person to advocate for a domestic SCC. Ever since the SCC appeared, the usual suspects—the fossil fuel industry, many Republican politicians, an array of conservative think tanks—have pushed for a domestic perspective. For example, in 2014, Representative Ann Wagner (R-MO) introduced the EPA Regulatory Domestic Benefit Act, which, she said, "would reveal the true costs of these harmful rules domestically on the American people." Perhaps because she represents people in a state that gets about three-quarters of its electricity from coal-fired power plants, Wagner also has been known to make inflammatory statements about most any attempt to reduce carbon emissions. During the previous administration Wagner often issued statements such as "[The Obama] administration is hell bent on waging a war on affordable energy and good-paying jobs."

To hear a more measured argument for using a domestic SCC, I met with Alan Viard, a resident scholar at the conservative American Enterprise Institute (AEI) who had provided written testimony on the issue to the OMB. In his testimony he argues that a global measure "is incompatible with the principle that the U.S. government exists to serve the American people, a premise that underlies federal policies in virtually all other areas. That principle calls for the federal government to consider only the costs and benefits experienced by the American people, an approach that requires the use of the domestic SCC." Assuming a situation in which the cost to America of eliminating a ton of CO_2 would be higher than the domestic SCC, Viard reasons using that global standard means America would be sacrificing a bit of its citizens' welfare in order to subsidize citizens of other countries.

When we met in his AEI office, Viard elaborated on this idea that using a global SCC is essentially a U.S. donation to other nations. He asked why the harm caused by climate change deserves special treatment. "There are hundreds of millions of people around the world suffering from [all sorts of] extreme deprivations today," he said, "but I don't think anybody is really proposing that the U.S. government has the same obligation to address their problems as it does the problems of its own citizens." He wasn't callous toward the tribulations of other nations, and he favored America providing its fair share of foreign aid. As he writes in his OMB testimony, "the lives and health of human beings outside the United States have no less moral value than the lives and health of human beings inside the United States." He just felt that the United States shouldn't unilaterally take disproportionate responsibility for the global costs of climate change at the expense of its own residents. By acting unilaterally, we enable nations that are dragging their feet on climate action to become free riders.

On the other hand, he told me, "If there was an international cooperative agreement, naturally it would make sense for all countries to agree to use the global cost measure." Viard spoke to me just before the adoption of the Paris climate agreement. That would be the same agreement that every nation in the world has signed on to, though Trump has started the wheels turning to withdraw the United States as soon as legally possible, which happens to be November 4, 2020, the day after the next presidential election. Now that a serious "international cooperative agreement" exists, what Viard considered a key impediment to the use of a global measure no longer exists. In fact, now that the Trump administration is no longer taking the actions to reduce emissions that the Obama administration embraced under the framework of the Paris accord, we Americans have become the free riders.

Paris aside, many advocates and scholars see other reasons for America to go global with the SCC. Most basically, restricting climate action to efforts that pass a cost-benefit test using a domestic SCC could gut climate change efforts. "If each country considered damages only within its own borders as a justification for action, then collectively, countries would fail to reduce greenhouse gas emissions enough to avoid the global damages that will ultimately affect every country," writes Karl Hausker, the World Resources Institute expert on the SCC. "Both textbook economics and common sense leads to this conclusion, a reality ignored by the EPA's new analysis."

Reciprocity is a key ingredient in mobilizing the international community to combat climate change. In 2016, Peter Howard and Jason Schwartz, two scholars at the Institute for Policy Integrity, published a paper on reciprocity and the debate over whether the United States should use a global or domestic SCC. In the paper they write, "Game theory predicts that one viable strategy for the United States to encourage other countries to think globally in setting their climate policies is for the United States to do the same, in a tit-for-tat, lead-by-example, or coalition-building dynamic. In fact, most other countries with climate policies already use a global social cost of carbon or set their carbon taxes or allowances at prices above their domestic-only costs. President Obama's administration has explicitly chosen to adopt a global social cost of carbon to foster continued reciprocity in other countries' climate policies." In sum, write Howard and Schwartz, "Only by accounting for the full damages of their greenhouse gas pollution will countries collectively select the efficient level of worldwide emissions reductions needed to secure the planet's common climate resources."

Applying Viard's pre-Paris testimony to the OMB, reciprocity is vital. "All countries could benefit from an agreement that each of them will curb all emissions that can be curbed at a cost lower than the global SCC," writes Viard. "If a properly implemented agreement of that kind existed, it would be desirable for the United States to comply with it by using the global SCC. In a cooperative framework, the losses that each country incurs from curbing its own emissions can be offset by the gains from the emissions curbs adopted by the other nations."

Our old friends externalities also play a part in justifying a global SCC. For example, when refineries process heavy oil, they are left with a dirty residue called petroleum coke, or "petcoke." Petcoke can be used as a cheap substitute for coal, but when burned, it emits even more toxic air pollutants and more carbon dioxide than coal does, so the EPA restricted its use on American soil. Thwarted in their own backyard, U.S. fossil fuel companies ramped up the export of petcoke to nations with looser regulations, with a large majority going to India. Already suffering from some of the foulest air in the world, India finally started imposing restrictions on petcoke imports in 2017, perhaps shamed by the sight of the visiting Sri Lankan cricket team wearing pollution masks during a big match with India. Umpires had to suspend play several times, and at one point a Sri Lankan player vomited on the field. India reacted to the immediate health effects of petcoke pollution, but the global warming impacts also constitute an externality exported to India from American producers, a harm that an American domestic SCC does not register.

Such obvious exports of climate damage aren't the only way America dumps global warming externalities onto foreign shores. Consider those shoes you're wearing, that smartphone in your pocket, or even the Barbie doll your daughter owns. Most likely that product was made in another country, though an American corporation may own the factory. It's even more likely if the item involves a pollution-heavy production process. Rich nations like the United States outsource much of their most greenhouse-gas-intensive manufacturing—we consume here at home, but some other nation does the dirty work. Yet when nations tally their emissions, China gets stuck with the tab for emissions related to the production of the Barbie doll.

The Global Carbon Project has tried to capture these transfers of responsibility for CO_2 emissions. An international coalition of scientific entities focused on climate change information, the Global Carbon Project did a study estimating nations' "consumption emissions" and their "production emissions." The project created a chart showing the flows from where CO_2 is generated to where the products associated with that CO_2 generation are consumed. By far the largest of the damning arrows points from China to the United States and Europe. But the use of a domestic SCC means the Trump administration is not accounting for these externalities.

At one point in my meeting with Viard our conversation turned to the national security repercussions of climate change and how they might influence the global-versus-domestic SCC debate. For years the American military has recognized and begun preparing for those repercussions. For example, in 2014 the Department of Defense (DoD) released its Climate Change Adaptation Roadmap (which, incidentally, the Trump administration has scrubbed from the DoD website). Speaking of the roadmap, then secretary of defense Chuck Hagel (a former Republican U.S. senator), said, "Among the future trends that will impact our national security is climate change. [Climate change] will intensify the challenges of global instability, hunger, poverty, and conflict." Obama routinely included the military aspects of global warming in his annual National Security Strategy document. Even Trump's first secretary of defense, James Mattis, has long advocated for the military to plan for the global disruption that climate change will cause. In written testimony to the Senate following his confirmation hearing, Mattis writes, "Climate change is impacting stability in areas of the world where our troops are operating today. It is appropriate for the Combatant Commands to incorporate drivers of instability that impact the security environment in their areas into their planning." But when Trump unveiled his 2018 National Security Strategy, climate change was missing from the list of global threats, which prompted more than one hundred members of the House of Representatives, including eleven Republicans, to send him a letter decrying this "significant step backwards."

I asked Viard if the fact that America's greenhouse gas emissions were contributing to droughts, extreme storms, sea level rise, and other destabilizing events around the globe that affect national security means that the United States should adopt a global SCC. He agreed that climate change poses a threat to national security, but he didn't see that as a reason to abandon a domestic measure. Instead he considered climate-related national security concerns a good example of climate damage to foreigners leading to climate-related damage to Americans, which would necessitate an upward recalculation of the domestic measure. Including such damages is "perfectly consistent with using a domestic measure," he said. "This measure could be augmented to include any cost that Americans would experience from those events. So those things would happen and harm Americans, in terms of national security or in some other way. A proper domestic measure would of course include those impacts."

I'm not sure how many other backers of a domestic SCC would see it Viard's way. National security costs are not the only climate damages that occur overseas and then bounce back to harm Americans. Most notably, given the interconnected nature of the global economy and the supply chains that bind the world, many of the injuries climate change could inflict on foreign economies would reverberate through the U.S. economy, too. A full accounting might result in a much higher domestic SCC that the fossil fuel industry and their supporters would find unacceptable.

Peter Howard and Jason Schwartz, who did the study on reciprocity, published a complementary report called "Foreign Action, Domestic Windfall" that tentatively quantifies the value to America of international climate action. Using the Obama-era SCC, Howard and Schwartz estimate that the steps to reduce greenhouse gas emissions already taken by other nations as of 2015 had produced upwards of $200 billion in direct benefits to the United States. If other nations follow through on the emission-reduction pledges they have made under UN efforts, such as the Paris accord, Howard and Schwartz figure that foreign actions will generate about $500 billion in benefits to Americans by 2030. And if other nations do their part in cutting emissions enough to keep the planet's average temperature rise below 2°C, by 2050 the benefits to the United States would climb to about $10 trillion. As Nobel laureate economist Kenneth Arrow wrote in his introduction to the study, "These are benefits which dwarf the higher energy costs of United States emission control. They are, in fact, almost surely conservative estimates of the benefits."

A 2018 study by five researchers from the United States and Italy took a different route but ended up in the same general vicinity as Howard and Schwartz: America will benefit if they and everyone else use a global SCC. The study looked at how much a ton of carbon dioxide emitted into the atmosphere will cost each of the world's nations. The authors calculated that the United States would incur extremely high costs, second only to India's. (However, some small countries would suffer proportionately greater damages.) This means the United States would benefit from emission reductions more than most other nations. As the lead author told *Inside Climate News*, "Our analysis demonstrates that the argument that the primary beneficiaries of reductions in carbon dioxide emissions would be other countries is a total myth." Incidentally, the researchers estimated that America's *domestic* SCC is $48, which is not only far higher than the paltry $1 to $6 suggested by the Trump administration but higher than the global SCC used by the Obama administration. The 2018 study produced a median global SCC of $417.

As with so many of the Trump administration's actions, its shift from a global to a domestic SCC may end up in court. One never knows what may happen at trial, but if the past is prologue, a global approach likely will prevail. In the highest-level legal test of the SCC so far, in 2015, the U.S. Court of Appeals for the Seventh Circuit heard *Zero Zone, Inc. v. United States Department of Energy*. The plaintiffs strenuously attacked the SCC, including its use of global benefits, employing the usual arguments. But the Seventh Circuit decided in favor of the defendants and ruled that the Department of Energy had "acted reasonably when it compared global benefits to national costs."

Even if there weren't plenty of other reasons to reject the America First attitude embodied in a domestic SCC, many people think the United States should adopt a global outlook because America owes the rest of the world. Home to about 4 percent of the world's population, the United States has emitted about 30 percent of the carbon dioxide that resides in our atmosphere and oceans. The United States has accounted for more emissions than any other nation, more than all the nations of the European Union combined. America's greenhouse gas history alone should compel the United States to pay much more than the domestic SCC indicates.

Even if there weren't plenty of reciprocity reasons to adopt a global outlook and even if the United States didn't owe the rest of the world due to its historical emissions, many people would say that altruistic concern for people beyond our borders should be reason enough for America to take strong climate action. When I was discussing the global-versus-domestic SCC debate with economist Kristen Sheeran, the climate change policy adviser to Oregon's governor, at one point she stepped aside from economic theory and lamented the fact that the usual suspects were fighting so hard to avoid helping folks elsewhere in the world. "God forbid we care about others," she said with a touch of wistful sarcasm.

7

EQUITY NOW

Climate change is taking Qaġġiġluilaq's home. By "home," I don't mean his house but his community and the land beneath it, the land on which he grew up, the land to which he and his community are bonded in a way that few residents of the modern world can fully grasp.

Qaġġiġluilaq is a twenty-one-year-old Inupiaq Eskimo who also goes by the name of Esau Sinnok. He's a climate lawsuit youth, serving as the lead plaintiff in *Sinnok et al. v. State of Alaska*, one of the state cases overseen by Our Children's Trust. He and fifteen other young Alaskans filed a climate suit against the state in 2011 that is still working its way through the courts. As with the federal lawsuit and the other state lawsuits, this case concerns future generations, but the ongoing destruction of Sinnok's home also illustrates the present-day plight of marginalized people around the world, whose lives are being disproportionately ravaged by climate change.

Now a student at the University of Alaska Fairbanks, Sinnok comes from the village of Shishmaref, tucked away on the northwest coast of Alaska just south of the Arctic Circle. This remote settlement of about six hundred people, almost all of them Alaska Natives, occupies the widest stretch of a narrow barrier island a few miles from the mainland. But that widest stretch measures only a few hundred yards across and grows narrower all the time.

The Arctic is warming much faster than most of the planet. According to the Alaska Climate Research Center, between 1949 and 2016 the annual mean temperature in Alaska rose 2°C. More important, winter temperatures have risen about 3.7°C, which means sea ice forms later in the year, melts earlier, and is not as thick or abundant as in the past. Without the sea ice as a barrier, storm surges and rising sea levels chew at the edges of Sinnok's home island, undercutting waterfront buildings and flooding parts of Shishmaref. Some houses have been lost, and about twenty have been moved away from the shoreline, but there's not much "away" left; in an average year the unimpeded ocean grinds ten or twelve feet off the island. During one monster storm the island surrendered about fifty feet of ground overnight. "Within the next two decades, the whole island will erode away completely," wrote Sinnok in an essay he posted in 2015.

Bending to the inevitable, in 1973, 2002, and again in 2016, the people of Shishmaref voted to try to relocate to the mainland, to someplace not far from their vanishing village. But it would cost about $180 million to establish a new village, far more than this poor community can afford. They've been seeking government support, but without success. Their hopes surged in 2016, when President Obama issued an executive order aimed at helping Alaskans adapt to climate change, but just a few months after entering the Oval Office, President Trump revoked that order. Trump's decision also bodes ill for the other thirty or so Alaskan villages grappling with the need to relocate due to an "imminent threat" from climate change, as a (pre-Trump) Government Accountability Office report put it. On a planetary scale, the UN estimates that fifty to two hundred million people, many of them already living on the metaphorical edge, will be displaced by global warming by 2050.

The residents of Shishmaref are trying to move to a nearby site because their traditional ways have evolved over millennia in concert with their surroundings. Elders pass down deep knowledge about subsistence pursuits, such as the timing of salmon runs, where and when to gather berries, the seasonal movements of caribou, and the best places to hunt for seals and walruses. Moving to a distant, largely unfamiliar environment would render much of this time-honored wisdom useless, which would not only make it hard for them to survive but would also unravel many of their cultural traditions.

However, even if the villagers somehow could secure enough funding to move, finding a suitable site nearby presents challenges because climate change is transforming the entire region. The loss of sea ice has reduced the populations of seals and walruses. Warmer winters have made it unsafe to store food in the villagers' traditional underground ice cellars. Warmer winters also have brought less snow and more rain, making it hard to journey by snow machine (snowmobile) to hunt and gather. In many places the permafrost is melting, causing buildings to tilt and crack. Instead of freezing solid all winter, what sea ice there is goes through cycles of freezing and thawing. Add in how thin the ice is and you have dangerous conditions for travel. A few years back an uncle of Sinnok's was hunting waterfowl in a place where in years past the ice had been safe at that time of year, but no longer was the ice thick and solid, and he fell through to his death.

Sinnok worries that his generation might be the last one to grow up in traditional ways. If the community can't relocate as a unit and its members are forced to scatter, they'd likely lose their unique dialect, their distinctive forms of art, their age-old stories, and many other cultural anchors. But even moving together to a nearby site won't enable the villagers to preserve most of their culture if global warming alters the area's environment so much that their largely subsistence lifestyle, so dependent on ice and snow, is no longer viable. If climate change warms the Arctic too much, Sinnok and his people simply may not be able to hold on to a way of life that they have treasured for thousands of years.

Shishmaref is an isolated village, but its struggle with climate change is not an isolated incident. Poor and disenfranchised people tend to be more vulnerable to global warming than their wealthier counterparts. For a sense of scale, consider that a 2013 World Bank survey estimated that about 770 million people live in dire poverty, defined as living on an average of $1.90 a day or less. (Depending on cost-of-living adjustments, three to five million Americans fall into this category.) Disparities in the ability to cope with climate change can exist for people living in the same city. For example, low-income neighborhoods may be located in areas that are more prone to climate-related flooding. This disproportionate vulnerability also exists among nations; many of the world's poorest countries lie in the tropics, which are being hit even harder by climate change than are nations in temperate zones. Compounding their climate-related hardships, impoverished neighborhoods and nations have less capacity to adapt to global warming.

Among the multitude of studies exploring the inequities associated with climate change is the UN's *World Economic and Social Survey 2016: Climate Change Resilience: An Opportunity for Reducing Inequalities.* "Sadly, the people at greater risk from climate hazards are the poor, the vulnerable and the marginalized who, in many cases, have been excluded from socioeconomic progress," writes then United Nations secretary-general Ban Ki-moon in the report. "We have no time to waste—and a great deal to gain—when it comes to addressing the socioeconomic inequalities that deepen poverty and leave people behind." The study estimates that over the last twenty years low-income countries have suffered a 5 percent drop in GDP due to climate-related disasters, while wealthy nations have not been smacked as hard. The World Bank also published a grim report on climate change and poverty in 2016. This study figures that global warming may push an additional hundred million people into poverty by 2030.

For years experts have been wrestling with whether and how to incorporate wealth disparities into climate economics and the social cost of carbon, though not always with equity as a goal. One early attempt showed how not to do it. As part of the Intergovernmental Panel on Climate Change's Second Assessment Report, produced in 1996, the working group on the economic and social dimensions of climate change tackled the always delicate task of putting a price tag on a human life. Despite dissent from some members, the group assigned different values to different lives depending on such factors as the average income of a person from a particular nation. Writing about the group's approach in their book, *Priceless*, economist Frank Ackerman and Georgetown University law professor Lisa Heinzerling report, "A careful reading of the fine print revealed that they were valuing lives in rich countries at $1,500,000, in middle-income countries at $300,000, and in the lowest-income countries at $100,000."

Understandably, this raised the hackles of many people, particularly residents of the $100,000 countries. They let it be known that they did not think that the life of, say, an Indian or a Nigerian was worth only one-fifteenth as much as the life of an American or a Saudi Arabian. The controversy dealt the IPCC Second Assessment a painful blow. When the Third Assessment came out five years later, it suggested a single value for everyone.

Karl Hausker, the SCC authority at the World Resources Institute, figures that the IPCC working group members who developed different values for different lives were not uncaring people, just neoclassical economists being neoclassical economists. Enveloped in a system that shuns ethics and assumes human welfare can be measured monetarily, they made what seemed like logical calculations. But using such a laboratory approach, said Hausker, means "you're going to get an answer that is ethically repulsive to everyone but economists."

Though the IWG didn't provoke an international incident, some of the same cold logic that haunted the IPCC Second Assessment cropped up in the IWG's calculations. For example, in one of the IAMs used by the working group the value placed on climate-caused deaths is adjusted according to the per capita income of different regions. The death of a twenty-five-year-old American was valued at more than $2,000,000, and the death of a twenty-five-year-old sub-Saharan African at about $35,000.

More broadly, the IWG simply swept equity under the rug. According to a paper by Michael Greenstone, the showrunner of the working group, and two coauthors, there are three reasons the IWG ignored equity. Two involve the technical difficulties of collecting and crunching the necessary data from around the world, and the third reason is that including equity in cost-benefit analysis violates standard operating procedure for the federal government. Observers debate whether these reasons hold water, but the fact remains that equity is missing from the analysis. "A well-established methodology for [including equity] was also available to the Working Group, but not used," write Laurie Johnson and Chris Hope in their 2012 paper. "With the majority of climate impacts expected to occur in low-income countries, this significantly lowered the Working Group's estimates." Perhaps standard operating procedure needs to be changed.

THE WEIGHT OF EQUITY

Economists concerned about the poor have developed tools that could bring the needs of the marginalized into the SCC equation. In particular, some scholars have applied a concept called "equity weighting" to the SCC and produced dramatically different results. For example, in the paper cited earlier, Johnson and Hope apply equity weighting to one of the IAMs and the SCC becomes up to twelve times larger, which would justify far higher levels of spending on reducing emissions.

Equity weighting stems from a fundamental economics concept: the law of diminishing marginal utility. Let's say you just baked a batch of chocolate chip cookies. Warm and enticing, they call to you, so you eat one. Heavenly! You eat another. Very tasty. You eat another. Pretty good. You eat another. Not bad. You eat another. It tastes okay, but you're getting a bit full. You eat another. The thrill is gone, and your stomach cries "no más." The more you have of something, the less satisfaction each additional unit provides—that's diminishing marginal utility.

Now let's assume you're one of the 770 million people in the world living on $1.90 a day or less. If your employer gives you a bonus of $1, it provides immense additional satisfaction, maybe enabling you to feed your kids twice that day instead of once. Now let's assume you're Jeff Bezos, Amazon's founder, who in 2018 enjoyed a net worth of about $160 billion and topped the list of the world's richest people, according to *Forbes* magazine. If you're fabulously wealthy and one of your investments earns you another dollar, it provides no additional satisfaction because what's one more cookie when you already have 160 billion?

Sustainability economists want to apply the law of diminishing marginal utility to climate economics and the SCC. They think global warming damage to a country like Haiti, with its per capita GDP of about $750 a year, should count for more in SCC calculations than the same amount of damage to a country like Norway, with its per capita GDP of about $75,000, one hundred times greater than Haiti's. Or, conversely, they think the benefits of reducing damages to Haiti should count for more than the benefits of reducing damages to Norway. The result would be a much higher global SCC.

Let's say an analysis finds that climate change will slice 10 percent off Haiti's gross national income, which would knock about $80 off the average Haitian's annual income. That means global warming would be costing this nation of eleven million people about $900 million a year—mere pocket change in terms of the economic impact of climate change on the global economy. But let's say the researchers estimating the SCC make an ethical decision that Haitians count as much as Norwegians, so they use a method of equity weighting that measures costs and benefits as a percentage of a person's income. For purposes of the SCC calculation, this boosts Haiti's annual climate change loss to about $90 billion—no longer mere pocket change. Now imagine applying the same method to all the 770 million people living in extreme poverty and to the more than two billion living in less extreme poverty. The result would be a much higher global SCC.

Without equity weighting or some similar mechanism, the SCC is indifferent as to whether climate change affects the rich or the poor; the impact on the net worth of consumption is all that counts. Let's say that by 2030 rising sea level damages an impoverished coastal village in Bangladesh to the tune of $5 million. Let's also say that the early years of warming lead to an increase in outdoor recreation in the United States, an expectation included in DICE. Let's further say a canny Wall Street investor anticipates this trend and plows money into a swanky golf resort in now-warmer Minnesota and rakes in $5 million. In the absence of equity considerations, the gains for the investor offset the losses for the Bangladeshis and all is well. Sure, the Bangladeshis would have spent their $5 million on food while the already wealthy investor spent his profits on a $5 million Louis Moinet Meteoris Collection wristwatch with pieces of meteorites embedded in its dial, but without equity weighting the SCC turns a blind eye to the distribution of costs and benefits.

James K. Boyce, professor emeritus of economics at the University of Massachusetts Amherst and a leading voice regarding equity and climate change, urges us to use some form of equity weighting when making decisions about the costs and benefits of emissions reductions. In addition, he urges us to apply the ethics that inform equity weighting when making decisions about spending money on adapting to climate change. We can protect more GDP by building seawalls to defend the beachside vacation mansions of the wealthy than by building seawalls to shelter poor coastal residents from the storms, but should we? "A different way to set adaptation priorities is to count each person equally, not each dollar," he writes. "This approach rests on the ethical principle that a healthy environment is a human right, not a commodity to be distributed on the basis of purchasing power, or a privilege to be distributed on the basis of political power." He adds, "In the years ahead, climate change will confront the world with hard choices: whether to protect as many dollars as possible, or to protect as many people as we can."

Calculating equity weights is complex and replete with value judgments, but sustainability economists think using the same standards to measure the impacts of climate change, whether they happen to Haiti or Norway, is also a value judgment. "There's this tendency to think that everything in economics is value-neutral because it looks so scientific and there are these very complicated mathematical, increasingly computerized models," said economist Kristen Sheeran. "But, of course, economics is a social science. Our models are models of societies. So every choice we make about how we're going to model society reflects, whether deliberately or not, a value judgment." She added that when economists say they're going to treat a dollar's worth of impact in Bangladesh and in the United States the same, it's a value judgment, and a telltale judgment to boot. She notes that the law of diminishing marginal utility shows up seemingly everywhere else in economics, but not when it comes to distribution and equity. She said that economists tend to like this law because it makes for nice demand curves on their graphs and, more seriously, it's one of the field's basic concepts. "But the one place [economists] don't hold that assumption is with respect to money income. In economics, it's never assumed that the more income you get, the less additional value each dollar conveys."

Other sustainability economists also point out this inconsistency, and some suspect it's a convenient oversight that serves the wealthy. If economists took the idea that a dollar means more to a poor Haitian than to Jeff Bezos and carried that idea to its logical conclusion, they'd have to acknowledge that redistributing dollars from the wealthy to the poor would increase societal welfare and therefore is economically efficient. As Herman Daly and Josh Farley put it in their textbook on ecological economics, some observers have accused neoclassical economists of omitting equity from their definitions of efficiency "precisely to sterilize the egalitarian implications of the law of diminishing marginal utility."

Peter Howard, the economist who heads the Cost of Carbon Pollution project, worries about the lack of equity measures in climate economics: "This is a big problem, actually.

If we continue on our current trajectory, it might be only a handful of people who get richer. Are we then really that much better off, or are a handful of people much, much better off?" Taking an even broader view, Amartya Sen, a philosopher and Nobel Prize–winning economist, sees the lack of equity as a profound flaw in the theoretical goal that neoclassical economics seeks to attain: the Pareto optimum. Markets achieve a Pareto optimum when they allocate goods and services such that no other distribution would make someone better off without making someone else worse off. But this says nothing about the initial distribution of wealth. The status quo may be one of immense inequality, such as exists in the world today. That means an economy is considered optimal even if 82 percent of the wealth produced in a year goes to the top 1 percent while the bottom 50 percent see no improvement—as was the case around the world in 2017, according to the nonprofit charity Oxfam. As Sen said, "A society can be Pareto optimal and still be perfectly disgusting."

The market does involve some implicit weighting, but it's not equity weighting. On the contrary, supply and demand puts its thumb on the scale in favor of the rich. Producers determine demand by observing which people are willing to pay how much for which products. Then the producers follow the money.

In their book, Daly and Farley cite the example of eflornithine, a drug discovered some four decades ago by the big pharmaceutical company Aventis. Aventis had been futilely trying to develop eflornithine as a cancer-fighting drug but instead stumbled onto its ability to cure sleeping sickness, a disease that at the time had reached epidemic proportions in many rural parts of Africa, infecting hundreds of thousands of people a year. The disease got its nickname because it disrupts sleeping patterns, but don't let the innocuous-sounding name fool you; sleeping sickness is a horrific illness. Victims usually suffer severe neurological damage and descend through a miasma of disorientation and madness until they die.

Effective and affordable, eflornithine seemed like the answer to the prayers of millions of Africans. However, it was affordable by the standards of a middle-class American but not by the standards of the poor Africans whom sleeping sickness typically infects. Recognizing that most of the people who needed eflornithine wouldn't be able to pay for it, Aventis ceased production. That's how business should be done, according to the neoclassical economics playbook.

The sleeping sickness–eflornithine scenario is not an anomaly. In the poor, hot regions of the world, such devastating diseases that could be affordably cured but go uncured are so common that they have an official term: "neglected tropical disease," or NTD. According to the Ending Neglected Diseases Fund, NTDs affect more than 1.5 billion impoverished people and kill some 170,000 a year. Note that a nonprofit like the Ending Neglected Diseases Fund would not have to exist if the market could address thorny equity issues such as NTDs, but it clearly cannot.

Luckily, the sleeping sickness–eflornithine story has a happy ending. Aventis licensed the drug to a couple of other companies, and they found out it can help remove unwanted facial hair in women. This cosmetic application spurred consumer demand in wealthy countries, so, with money to be made, the companies revived the production of eflornithine. Of course, the market steered the drug to the treatment of facial hair, but the profits from that market enabled the companies to provide eflornithine at reduced prices or for free to Africans with sleeping sickness. At least, they provided it after pressure from nonprofits and activists. Not exactly a testimonial to the magic of the market, but millions of people did get the cure they needed.

The needs of the poor also get lost due to the methods

economists sometimes use in cost-benefit analysis (CBA). When policy decisions involve nonmarket goods, economists sometimes conduct willingness-to-pay (WTP) surveys to estimate the value of, say, preserving a free-flowing river instead of damming it. So they ask a varied selection of people who would be affected by the dam how much each would be willing to pay to keep that river running wild. Perhaps the river runs through an undeveloped area populated by people leading a largely subsistence lifestyle who depend on that free-flowing river for almost everything. However, they're money-poor, and when asked how much money they would spend to keep the river undammed, their impoverished perspective causes them to suggest very low amounts. On the other hand, when researchers ask some upper-middle-class guy in a nearby city, who would benefit from the electricity the dam would produce, how much he'd be willing to pay to dam the river, he might say that he's willing to spend hundreds of dollars because that's a pittance to him even though it's a year's income for the river dwellers. Matthew Kotchen, a CBA expert and economics professor at Yale, writes, "Measuring costs and benefits with WTP is also objectionable to some because it depends so heavily on the distribution of income. People with lower incomes have a lower ability to pay and therefore have less influence on the outcome of a CBA."

My hypothetical free-flowing river broaches an entire category of goods and services crucial to the lives of poor people that is being taken from them partly due to their inability to pay: ecosystem services and the goods that nature provides. Many of the world's destitute rely heavily on the free provisions bestowed by the natural world. They suffer disproportionately from such problems as biodiversity loss, polluted air and water, overfishing, and climate change. Using market mechanisms, such as a carbon tax informed by an SCC, to internalize some of those externalities can help stem the environmental harm. However, the use of such mechanisms complicates the allocation of essential ecosystem goods and services for poor people due to their lack of a voice amid a supply-and-demand din in which the volume of one's voice depends on the volume of one's bank account.

In a 2014 study, a quartet of researchers from the United States and Brazil examined the equity issues inherent in market-based approaches. Citing a lack of large-scale examples of market-based instruments dealing with essential ecosystem services, the researchers decided they needed a proxy, so they examined the market allocation of another essential resource: food. "Obviously there's no more important economic sector than food," said Josh Farley, one of the quartet. "We die if we don't have it."

Part of the study looked at how the allocation of food can be affected by global shortages of basic agricultural commodities and the resulting price surges for food, as happened in 2007–2008 and again in 2010–2011. Such surges serve as proxies for the kinds of price increases that would result from internalizing externalities and from the rise in prices for nature's goods that happens as they become scarcer. The researchers contrast the United States with Zambia. The study points out that Americans consume an average of 3,750 calories a day, that they have the highest recorded rates of obesity in history, that they waste about 40 percent of their food, and that they spend only about 7 percent of their income on food. By contrast, average Zambians spend about 60 percent of their income on food but eat only about half as many calories as Americans, considerably less than the recommended amount. The authors add that malnutrition is by far the leading cause of death for Zambian children.

When shortages caused global prices for rice and corn to double in 2007–2008, Americans felt a little pinch as prices at the supermarket rose about 18 percent. But for most Americans, even that unusually steep price increase didn't change what or how much they ate because they could easily bump up the percentage of their relatively large incomes that they spent for food. Research indicates that food consumption in the United States during those rough agricultural years declined only 1 or 2 percent. Zambia was a different story. Average Zambians were forced to spend even more than 60 percent of their income on food, sometimes approaching 100 percent; food is a need, not a want. But when you start at 60 percent of an already rock-bottom income, there's not much room to boost your spending. "With an essential resource like food, when people are barely getting enough anyway and the price goes up, they want just as much," Farley told me. "Their ability to buy it goes down, but their physiological demand doesn't go down." He added, "I'd say that to allocate food according to willingness to pay instead of physiological need leads to obesity and starvation at the same time." In the end, the market directed most of the food to wealthier nations, and poor, hungry people just got poorer and hungrier.

The same thing happened in the food price surge of 2010–2011. "You and I didn't eat a single slice of bread less," said Farley. "In rich countries there was no reduction of staple grains, but in poor countries forty million more people went hungry." Farley said that the food shortages also led to rioting and political unrest. "The market systematically allocates food to the overfed instead of to the hungry. When there's a food shortage, the only people who reduce consumption are the starving. Markets, regarding the allocation of essential resources like food, are completely perverse."

Having found empirical evidence of the market's innate inability to equitably distribute food, the quartet of researchers looked at ecosystem services through the same lens. As context, they point out that markets do a poor job in general of allocating public goods, which include many of the most vital ecosystem services. Still, market approaches can play useful roles if they're properly designed, and the quartet argues that proper design should include giving equity an important seat at the table.

The study uses the example of a forest. As we saw in chapter two, a forest provides many free benefits for people, such as the reduction of soil erosion along streams, the absorption of rainfall and subsequent slow release of water later during the dry months, the filtration of pollutants from runoff, the moderation of floods, and the provision of habitat for game animals. Poor people often need these free services; they can't just run out to the store to buy a rib eye if there is no game to hunt. But the locals don't have enough in their wallets to influence the allocation of the forest's goods and services, so they would lose out if this forest were integrated into a market. The market likely would scalp that forest to produce lumber. "We're using market mechanisms to determine the value of resources provided freely by nature, but market mechanisms are determined by one dollar, one vote," said Farley. "That's what markets are. They're the antithesis of democracy. They're absolutely plutocratic. You have no right to vote in a market unless you have money. When trying to decide what the value is of ecological processes inherited from nature, it seems to me one person, one vote makes a lot more sense than one dollar, one vote."

Do equity concerns undermine efforts, like the SCC, to calculate the full costs of environmental externalities and incorporate them into prices? That depends on whether and how those calculations take equity into account. It also depends on what we do after we make such calculations and use them to raise prices to reflect social costs. If we want to be fair, we'll have to do more than simply determine new prices and then leave it to the market to sort things out. Farley cites the example of industrial agriculture, which produces huge climate change externalities. "If we tried to internalize eco-logical costs into food prices, it would send those food prices skyrocketing. You and I wouldn't notice, but poor people would starve in droves. So this idea of internalizing ecological costs is really, really tricky when you live on a wildly unequal planet." Still, he thinks internalizing omitted costs can help protect the environment, but at the same time we need to employ ethics and wisdom, not just supply and demand. Yes, there are hurdles, says Farley, "but that doesn't mean it's a bad idea. It just means you have to pay very close attention to the impacts on poor people, as well."

8

INTERNAL MEDICINE

Josh Farley got it right when he said that internalizing is really, really tricky. He was speaking in the context of equity and the environment, but internalizing in general is difficult to pull off. Yes, internalizing can play an important part in reducing social costs, but some economists mistakenly see it as close to a panacea. This overestimation comes particularly from conservative econs, some of whom see internalizing as a market-oriented strategy that can restore the simple beauty of supply and demand; just put a price tag on externalities and once again producers and consumers will receive accurate price signals so they can make decisions based on the full costs of things, thus optimally allocating resources.

In the realm of global warming, one way to internalize social costs is to estimate those costs in dollars (perhaps using the SCC process) and then set a price based on that estimate. As we'll see below, some people are advocating for just such an approach, recommending that we take the last pre-Trump SCC (about $40) and charge large producers of emissions $40 for every ton they release. But using the SCC as a guide is not the only method. Some governments have set much lower carbon prices, choosing a number that seems politically palatable rather than one intended to compensate for externalities. A few governments have set higher carbon prices, choosing a number that the decision-makers hope is high enough to spur significant changes in consumer behavior.

In the United States many liberals have long advocated internalizing the social costs of carbon via a price on carbon emissions. So have multitudes of economists; in 2019 more than 3,500 economists signed on to a statement published in *The Wall Street Journal* supporting a carbon tax, though one with some of the problematic features we'll discuss below. On the other hand, almost all of the nation's conservative leaders, even those few who have publicly acknowledged the reality of global warming, have long denounced carbon pricing as an unnecessary tax. However, in the last few years, small but increasing numbers of laissez-faire disciples have warmed to carbon pricing, though their approaches and motivations differ in many ways from those of the political left.

Avid proponents of carbon pricing, particularly the handful of suddenly zealous conservatives, assert that the right carbon pricing plan would be the main tool—maybe even the only tool—needed to solve global warming. This confidence in internalizing lies at the heart of a 2017 proposal, led by a group of conservative luminaries and supported by some oil and gas companies, to address climate change with a carbon tax (which they refer to as a carbon fee).

This group calls itself the Climate Leadership Council (CLC). In 2018, an organization named Americans for Carbon Dividends (AFCD) formed to lobby for the proposal, which they tout as the right-wing way to deal with climate change. As AFCD puts it, "This plan is based on the conservative principles of free markets and limited government."

I question the motives of some of the key drivers of CLC and AFCD, like former senators Trent Lott and John Breaux, the co-chairs of the AFCD. Both dutifully supported the fossil fuel industry while in office, and both quickly scooted through the revolving door and became big-time lobbyists after leaving office, with clients that include fossil fuel companies. Perhaps they sincerely worry about climate change and just want to do their part to reduce emissions, but their history suggests they also harbor other intentions.

I'm even more dubious about the motives of some of the corporate members of the CLC and AFCD, notably Exxon-Mobil, Shell, BP, and the other fossil fuel companies; I'm confident they're mostly interested in profits, not reducing emissions. It so happens that the initial CLC plan offers goodies meant to please fossil fuel interests. For example, the plan proposes "the elimination of regulations that are no longer necessary upon the enactment of a rising carbon tax." The plan further states that "much of EPA's regulatory authority over carbon dioxide emissions would be phased out, including an outright repeal of the Clean Power Plan." The very next sentence in the CLC's description of their plan serves up another tasty treat: "Robust carbon taxes would also make possible an end to federal and state tort liability for emitters." Meaning, let's make a deal that would allow the fossil fuel companies to elude the rising wave of climate change lawsuits.

But not all the CLC plan's supporters are corporate lobbyists or conservative ideologues. Remember that enthusiasm for internalizing the costs of carbon comes mostly from liberals. A smattering of moderate Democrats, credible scientists, mainstream economists, and environmental groups also backs the CLC initiative, at least as a starting point for discussions. With caveats, even James K. Boyce sees much to like about the CLC effort.

Boyce is a leading progressive economist with a long track record as a champion of forceful and equitable climate action.

His 2019 book *The Case for Carbon Dividends* lays out a plan that shares two key elements with the CLC approach. One, Boyce agrees with the fundamental proposition that a carbon price can significantly help move consumer and producer behavior toward clean energy. Two, he likes the CLC idea of paying dividends. The CLC proposal favors a dividend plan that would channel most or all of the money generated by the carbon price equally to all Americans.

Boyce sees several benefits to dividends. If done with equity in mind, it could provide low-income Americans with enough money to make up for the higher prices on various goods and services caused by internalizing the social costs of emissions. In his book, Boyce examines a hypothetical plan that features a carbon price of $100 a ton and returns 100 percent of the proceeds in equal amounts to every American. He estimates that such a plan would boost the income of 80 percent of Americans. Only the richest 20 percent would experience a net loss—generally about 1 percent of their ample incomes— because they typically buy more taxable, fossil-fuel-related goods and services than low- and middle-income consumers. Boyce further likes dividends because he thinks they would help generate political support for a carbon price.

However, when Boyce and I spoke, he expressed a number of reservations about the CLC's plan, such as its intent to start with a price of $40 per ton. "Is $40 a ton enough? No. Is it better than what we have today? Yes." He thinks the price may need to be much higher. Boyce advocates price hikes at regular intervals if a carbon price plan isn't lowering greenhouse gas emissions fast enough to meet a serious target, like that of the Paris accord. The CLC plan includes annual price hikes, but at the feeble rate of 5 percent.

Most important, according to Boyce, the price must be anchored to a science-based emission-reduction trajectory, rather than follow the usual method of just setting a price and hoping for the best. Boyce sees mandatory limits on emissions as the litmus test for any serious carbon pricing policy. The CLC may have passed that test in the fall of 2019 when it beefed up its proposal with what it calls an "emissions assurance mechanism." This mechanism is supposed to accelerate the annual rises in the carbon price as needed to meet the CLC's emission reduction targets. However, the plan remains short on detail, and this key provision would need to be rock-solid and free of loopholes to provide the assurance promised by the term "emissions assurance mechanism." And, of course, the targets would have to be sufficient.

Boyce also has misgivings about some of those goodies the CLC originally offered to fossil fuel interests, notably the one absolving them of legal liability. "To me the single most objectionable thing that they slipped [into their original document] is that you can have a waiver for tort liability for fossil fuel corporations. So there would be some sort of legal exemption that would be accorded them that would protect them from the damages they've done." In Boyce's view, retaining the right to sue fossil fuel companies is both a moral imperative that enables people to seek just compensation and a lever for pressing the industry to clean up its act. As the industry well knows, legal actions are not a marginal matter. Research from Columbia Law School and the Arnold & Porter law firm finds that worldwide more than 1,300 lawsuits related to climate change have been filed, most of them in recent years and many of them aimed at fossil fuel businesses. Under pressure, late in 2019 the CLC indicated that it would drop its proposal to shield industry from lawsuits, but that could be a deal-breaker for some business interests. Whether protection from legal liability is permanently banished from the proposal remains to be seen.

Many climate advocates, including Boyce, also have misgivings about the proposal's aim to eliminate regulations "that are no longer necessary upon the enactment of a rising carbon tax." Sure, theoretically a price on carbon could obviate the need for many climate regulations if the price rises as high and rapidly as necessary to meet stringent science-based emission reduction targets. But should we trust that this ideal result will happen in the real world? Defining when regulations "are no longer necessary" could become a fraught, politicized tussle. Boyce also calls attention to some of the benefits that regulation could deliver even if a price alone resulted in sufficient carbon reductions, such as rules aimed at limiting other greenhouse gases.

The Energy Innovation and Carbon Dividend Act takes a crack at the trust problem. This proposed legislation is generally rigorous and nominally bipartisan, though only one of its seventy-five co-sponsors is a Republican, and he's retiring after the 2020 elections. Essentially, one provision of the bill suspends climate regulations for ten years for the industries taxed under the act. If after ten years the emissions targets are not reached, the EPA is directed to once again exercise its regulatory muscle and steer those industries as necessary to meet the targets.

This act's approach beats simply ditching climate regulations, but that ten-year gap carries risks. Consider the IPCC special report that rocked the climate world in 2018. It emphasized that the twelve years between the report's publication and 2030 would be crucial. The report finds that we must reduce emissions to about 45 percent below 2010 levels by 2030 to have a good shot at holding the temperature rise at or below the greatly desired 1.5°C mark. Given the gravity of the coming decade, we might do better to adopt a regulatory strategy such as that included in the bill championed by Senator Sheldon Whitehouse, which retains climate regulations along with instituting a carbon price. If the carbon price in fact makes climate regulations "no longer necessary," that would be great, but let's see the proof before we relax the rules.

A 2019 study published in the *Proceedings of the National Academy of Sciences* underscores the urgency of the next decade and the need for any carbon tax plan to absolutely ensure that emissions decline swiftly, no dithering allowed. Authors Kent D. Daniel, Robert B. Litterman, and Gernot Wagner recommend that we start with a high carbon tax (maybe in the $100-to-$200 range) and then let that tax decrease. This flips the usual pattern of starting with a low tax and raising it over time, but the authors emphasize the need to make things happen quickly. Then, as the tax-stimulated changes and innovations bring down emissions (assuming the tax works as advertised) and as the costs and risks of climate change diminish, the amount of the tax can also diminish. The authors also calculate that the costs of delaying an aggressive tax would be punishing: an extra $1 trillion for a one-year delay, an extra $24 trillion for a five-year delay, and an extra $100 trillion for a ten-year delay.

We would be foolhardy to gamble the planet's future by putting all our eggs in the basket of carbon pricing, especially carbon pricing not tied firmly to a science-based, mandatory-emissions-reduction trajectory. Given conservative politicians' and the fossil fuel industry's track record of obstruction, deception, and delay regarding climate change, we should remain wary of their power to undermine the design and implementation of carbon pricing efforts. But this doesn't mean we should dismiss every aspect of their plans, and it certainly doesn't mean we should not include a carbon price among the instruments we use to fix climate change. After all, a carbon price is the kind of systemic change we need, a solution that pervades the economy. We would be wise to consider the strengths and weaknesses of all the plans being floated and use that knowledge to fashion the best carbon pricing approach possible.

Sadly, all this discussion about a carbon tax may ultimately be just so much hot air.

The conventional wisdom says that neither the CLC proposal nor any other carbon pricing plan stands a chance of becoming federal law in America given the current political climate. The conservative leaders of the CLC and AFCD are mostly old-school former politicians whose views represent only a sliver of the contemporary Republican Party. Today's GOP, dominated by climate-denying, tax-hating Trump supporters, seems unlikely to support a carbon tax no matter how much its backers try to align it with conservative principles. To make their feelings clear, in 2018 the then-Republican-led House of Representatives passed a resolution condemning a carbon tax, calling it "detrimental to the United States economy." Two hundred twenty-two GOP members voted for the resolution, and only six voted against it. Even in other nations whose federal governments are more open to climate action than the United States, carbon pricing has often sputtered. Some forty countries have enacted carbon prices, but seldom has the impact been meaningful, and never has a carbon price achieved anywhere near the needed emission reductions.

The resistance to carbon pricing remains strong in America, but starting in 2019, contrary murmurs could be heard. For example, the U.S. Conference of Mayors urged Congress to put a price on carbon that would reduce emissions in line with the Paris accord. Also in 2019 and even more so in 2020, a few current GOP members of Congress began openly discussing conservative approaches to climate policy, with a carbon price in a starring role. One motivation driving these Republican discussions seemed to be a desire to head off any Democratic climate initiatives, an understandable worry given the growing number of centrist and progressive proposals for addressing climate change, the numerous polls showing the public's rising appetite for climate action, the flurry of climate change lawsuits, and the outpouring of local and state climate change regulations. Increasingly, the thin red line of the largely GOP-controlled federal government seems like the main barrier that stands between industry and serious climate regulation. Or, perhaps, between industry and a carbon pricing plan with a decisively higher price and no goodies. Knowing that the wall is weakening and could crumble after the 2020 elections, some fossil fuel interests and a handful of conservative politicians apparently have concluded that some kind of major climate action is likely and that a deal like the CLC plan may be the best they can hope for. If conservative climate proposals do gain traction, the concept of internalizing may become pivotal to the future of climate action in America.

So how far can internalizing take us down the road to a stable climate? More broadly, when dealing with environmental and social issues in general, how far can we get by internalizing costs and using the market?

The Rhodium Group, a leading independent research firm that often examines climate issues, published a report that includes insights into some of the limitations of internalizing climate costs. It conveyed a key general point: not all producer and consumer decisions are equally responsive to a carbon price.

As one example, Rhodium examines the concept of the ratio of higher capital cost to operating cost and applies it to passenger cars. Buying a car (the capital cost) is far more expensive than buying gas (an operating cost), but the carbon taxes being proposed would mainly affect the gas price and barely touch a car's sticker price. As history has shown, raising the price at the pump won't do much to deter someone from driving when that someone has spent tens of thousands of dollars to purchase a car, not unless we hike the price of gas enormously. And no one with political power is suggesting a carbon price anywhere near high enough to cause that kind of price hike.

On the other hand, a carbon tax can hit hard enough to induce changes when levied on businesses with high operating costs that emit large amounts of greenhouse gases. Coal-fired power plants provide the prime example. Existing plants require few additional capital expenditures, but they continually spend a huge chunk of their budgets on coal, so if a tax made that coal significantly more expensive, then utilities would look for a cleaner fuel. Fortunately, numerous cleaner fuels are available; having ready alternatives is another element that makes some industries more responsive to carbon pricing than others. Returning to our car example, the lack of an easy alternative to gasoline is another reason driving is largely immune to carbon pricing. The Rhodium study finds that by 2030 United States coal production would fall 28 to 84 percent compared to current policy, depending on the size of the carbon tax, while during the same time period transportation emissions would only drop about 2 percent given a tax of $50 a ton.

Carbon pricing also suffers from limitations related to economic fundamentals. Longtime climate change leader Hal Harvey and some colleagues from Energy Innovation, the energy policy firm he heads, touch on a few of those intrinsic shortcomings in their book, *Designing Climate Solutions*. They write about "economic signals," which include both carbon prices to discourage pollution and financial incentives to encourage energy-efficient behavior and products. They note that "well-known market failures and transaction barriers restrict the ability of economic signals to encourage adoption of low-cost—or even cost-saving—energy efficiency upgrades that would reduce emissions."

Among the examples Harvey and his coauthors cite are "split incentives," as evidenced by most rental properties, "when renters, and not building owners, who typically make the capital investment decisions that affect energy efficiency, pay the energy bills. A landlord not paying utility bills has little reason to upgrade an apartment fitted with an inefficient water heater and refrigerator, but the renter is in no position to make a capital improvement on the building. The economic opportunity is missed, and economic signals alone won't fix it. In contrast, a good building code . . . can get the job done."

Given the complex strengths and weaknesses of carbon pricing, plus all the design decisions and assumptions that go into the models, it's tough to calculate how big a dent a price could make in greenhouse gas emissions. Harvey and his coauthors give it a try. Based on a carbon price that rises to $50 a ton by 2030 and increases 2 percent annually after that, they tentatively estimate a reduction of at least 26 percent of what's needed to keep the global temperature increase below 2°C. Based on a similar price of $50 a ton rising at an inflation-adjusted rate of 2 percent a year, the Rhodium report figures we could achieve emission reductions of 39 to 47 percent below 2005 levels by 2030. However, those percentages result largely from picking the low-hanging fruit, notably closing coal-fired power plants, so the benefits of typical pricing plans would diminish in the longer term.

Much uncertainty colors such estimates, but Harvey and his coauthors and the researchers at Rhodium feel quite certain about one thing: carbon pricing alone will not suffice. As the Harvey team puts it, "Carbon pricing is not a silver bullet to achieve the deep emission reductions needed . . . Rather, it is one important part of a package of policies." To leave no doubt about their feelings, they add, "Many economists suggest that carbon pricing, set at the social cost of carbon . . . is the only policy needed to efficiently reduce emissions. This assertion is false. However, carbon pricing is an essential part of a policy portfolio for tackling emissions."

Boyce thinks a carbon price could have a positive impact even greater than what Harvey and Rhodium estimate, but only if we adopt an approach that puts biophysical reality in the driver's seat. Typical plans (like CLC's) set a carbon price and then let the economy react to that price, thus leaving it to supply and demand to determine how much society reduces emissions. But what if that price is wrong?

In this book we have discussed at length the mental gymnastics required in the making of IAMs, the SCC, and, now, the price for carbon. Our economic gymnasts must come to grips with assumptions, omissions, spotty data, equity concerns, attempts to price the priceless, and layer upon layer of complexity. So far none of the gymnasts has delivered anything close to a perfect performance, which is hardly surprising given the degree of difficulty. Uncertainty inhabits every twist and turn. So Boyce would have us keep our eye on the prize: the amount of emissions reductions that science and our values indicate we need. Instead of setting a price and letting the quantity of emissions adjust to that price, Boyce recommends setting targets for emissions reductions and letting the price adjust as needed to meet those targets. Then the price would climb as high as necessary. Strictly speaking, this wouldn't be internalizing social costs into market prices so much as using carbon prices as a regulatory tool to achieve a social goal. Boyce points out that society also might decide to keep carbon prices lower by liberally using policies that complement a carbon price, such as regulation and public investments designed to spur innovation.

As noted above, even with all the time and effort that has been devoted to determining the costs of the externalities associated with climate change, considerable uncertainty taints the numbers. That lack of assurance also is a good reason to question the notion of widespread internalizing as a realistic answer to social costs in general. But it's not just the uncertainty. The fact that it has taken so much time and effort by so many people to hash out even halfway-decent estimates of the SCC and carbon prices leads to a general truth about internalizing: it's hard to come up with a good number. Yet supply and demand can't kick into gear without a number. This is a problem.

If internalizing were going to enable the market to save the world, the economy would have to price and internalize externalities from a multitude of sources: biodiversity loss, overfishing, water pollution, urban sprawl, topsoil depletion, plastic waste, excessive logging, wetland destruction, and on and on; the list is literally endless. And those are only the environmental externalities. As a practical matter, how do we price and internalize all these costs?

Take the example of consumer and industrial chemicals, many of which can damage the environment, some of which can make us sick, and a few of which can kill us. To internalize their social costs we would first need to know which chemicals are toxic. Good luck with that. About eighty thousand chemicals have been used in commercial applications in the United States, and we have little idea how many are harmful. Under the original chemical safety law, industry didn't have to prove that a chemical is safe before selling it; the EPA had to prove it *wasn't* safe in order to take it off the market. This reversal of the rational burden of proof was bad enough. But intense lobbying by industry led to a catch-22 situation in which the EPA couldn't require a company to test a chemical for risk unless the agency could first show that there's good reason to think the chemical poses a risk, which is tough to do without testing. No wonder less than 1 percent of chemicals in American products have been thoroughly evaluated for toxicity. Even after the EPA identifies a chemical as toxic, it's extremely hard to actually ban it.

After decades of inaction, in 2016 Congress actually managed to pass a bipartisan bill that offered significant improvements to the feckless legislation that had been regulating chemicals. However, the Trump administration has changed how the EPA is carrying out the 2016 law, hugely reducing its effectiveness.

Even if the EPA had been doing its job diligently for decades, the percentage of chemicals tested would still be low. It takes a lot of time and money to test a chemical, and one thousand to two thousand new ones enter the market every year. Keeping up would be very hard and would require a tremendous amount of resources. But if the EPA doesn't catch up and keep up, we'll continue to have loads of untested chemicals. And testing is only the first step in determining the social cost of toxic chemicals. If the EPA finds that a chemical is hazardous, they then need to figure out how hazardous. Does it kill people? Does it make people sick? Does it harm plants and animals? How many people does it kill and sicken? How many plants and animals does it harm? And even if they can answer these and the many similar questions, the EPA would have to quantify all the costs in order to get a usable number.

Now extrapolate from chemicals to all the other products and services and activities that produce externalities. Ironically, the only way to even try to do all this internalizing would be to establish a gargantuan government bureaucracy whose calculations would exert vast control over the market. Not exactly the small-government vision celebrated in song by the Milton Friedman Choir.

But fear not. No Bureau of Internalization will ever cast a looming shadow across our nation. Such comprehensive internalization is plainly impractical. Even if it were technically possible, it would be prohibitively expensive. Instead, in the vast majority of cases, government will need to exercise its judgment and regulate market missteps—which means we're going to need a government with good judgment.

In a few cases internalizing can still play a vital role, with a carbon tax being exhibit A. All the work on IAMs and SCCs can help set a price that isn't entirely arbitrary. Better yet, institute a more robust SCC effort and set a more accurate carbon price, one that better accounts for social costs and risk factors that are hard to quantify but that we, in our collective wisdom, feel we should do our best to estimate and include. How much higher? Well, the 2018 IPCC report recommends a per-ton carbon price between $135 and $5,500 by 2030 and a price between $690 and $27,000 by 2100. (No, all those zeros are not typos.)

We could afford to bankroll group efforts to roughly estimate the social costs for a select number of pressing issues, but that would be just a drop in the social cost bucket. Besides, some sustainability economists contend that internalizing suffers from drawbacks that go beyond the infeasibility of handling the sheer volume of externalities.

In their book, *Ecological Economics*, authors Herman Daly and Josh Farley offer some valuable insight. They suggest that the issue of internalizing takes us back to the basic purpose of an economy: the allocation of limited resources among competing ends. They doubt the ability of the market, through internalizing, to wisely allocate all, or even most, of what matters to us. "Price provides a feedback mechanism used by the market to maximize profit, which economists assume creates the appropriate conditions for maximizing human well-being," they write. "We . . . cannot support the reductionist approach of assuming that the profit motive alone is sufficient to maximize human well-being, much less to guide us in any quest toward an ultimate end."

Daly and Farley think people make a mistake when they assert that they can quantify and internalize costs associated with certain nonmarket goods that Daly and Farley consider categorically different from market goods. The authors mention our right to vote in a democracy. "We could readily devise a survey that would tell us how much a voter would be willing to pay for the right to vote (or alternatively the minimum amount for which a voter would sell her vote). We could do the same for human rights, and many people consider the right to live in a healthy environment such a right. But most people would probably agree that politics and human rights are in a different moral sphere than economics, and power in the sphere of economics should not translate to power in these other spheres." They add, "Putting dollar values on everything does not make the necessary decisions more objective; it simply obscures the ethical decisions needed to make those 'objective' valuations."

9

GROWING PAINS

Okay, so internalizing won't solve all our problems. Not to worry. Orthodox economics offers another all-purpose solution, one it trots out much more often than internalizing: Grow. Grow the GDP. Grow new businesses. Grow trade. Grow stock prices. Whatever the problem—pollution, inequality, unemployment, crime—conventional economic growth supposedly will alleviate it or even solve it altogether.

Sustainability economics holds a more complicated view of conventional economic growth, which we'll dive into in a few pages. But first let's consider a forbidding question that lurks in the SCC research: Will climate change, if not aggressively addressed, derail conventional economic growth?

On September 20, 2017, Hurricane Maria howled onto the shores of Puerto Rico. Revved up by climate change, the category four behemoth rampaged across the island, killing thousands of people and flattening buildings, trees, power lines, and most everything else in its path. In the immediate aftermath, relief efforts and public attention understandably focused on the tribulations of the island's 3.4 million inhabitants. Few people were thinking about business as suddenly homeless Puerto Ricans waded through knee-deep floodwaters and jammed onto cots in sports arenas, struggling just to find food, water, and shelter. But with the economy as broken as all those snapped palm trees, Puerto Rico needed to get the wheels of commerce turning again as soon as possible.

Unfortunately, it could take a generation for Puerto Rico merely to bring its economy back to where it was before Maria blasted the island, according to estimates made by Solomon Hsiang and Trevor Houser. Hsiang is an associate professor at the University of California, Berkeley, and an expert on the relationship between climate change and economic development. Houser heads the Rhodium Group's energy and climate team and co-directs the Climate Impact Lab. Interested in the scope of economic damages caused by climate-related disasters, Hsiang and Houser conducted a study of Puerto Rico's post-hurricane economic prospects and came up with the gloomy estimate that it would take twenty-six years for the per capita income of Puerto Ricans to return to pre-Maria levels. As Hsiang and Houser put it, "Every sixty seconds that Puerto Rico suffered through Hurricane Maria, roughly a week's worth of economic development was lost." By way of comparison, they point out that the Great Recession of 2007–2009 reduced America's economic output by 9 percent whereas the harm wrought by Maria has the potential, if the island doesn't get abundant aid, to constrict Puerto Rico's output by 21 percent. That amounts to a cumulative loss of about $180 billion.

The bleak future of Puerto Rico's post-Maria economy takes us back to a pivotal subject we touched on in an earlier chapter: the fantasy that even an extreme warming of the planet would not noticeably affect the global economy because inexorable economic growth will overwhelm any impacts stemming from climate change. You may recall Martin Weitzman's calculation in which he extrapolated from a leading IAM to show how it grossly underestimated the economic impact of a 10°C global temperature rise over two centuries, assuming an annual growth rate of 2 percent. According to Weitzman, the dubious result indicated that such an extraordinary temperature increase would merely cause the economy to grow only forty-four times larger rather than fifty-five times larger during those two hundred years. In other words, even catastrophic climate change would supposedly make our descendants only slightly less rich than they would be without any global warming, in which case strict cost-benefit logic tells us not to bother spending money now on reducing emissions because future generations will be wallowing in wealth no matter what we do about greenhouse gases.

Many studies, however, including Hsiang and Houser's, undermine the view that growth will march on, barely slowed by warming, and heal all ills. Let's return to post-Maria Puerto Rico. Hsiang and Houser researched the impacts on growth from a single hurricane, albeit a whopper, but more hurricanes loom in the island's future. (Not to mention Hurricane Irma, which hit Puerto Rico just two weeks before Maria.) Those storms will be increasingly intense as water and air temperatures keep climbing. Drought, heat waves, sea level rise, ocean acidification, and other climate-related harms may join hurricanes in repeatedly stunting Puerto Rico's growth. It's impossible to know how much of a beating the island's long-term economic growth will take if the world doesn't reduce its emissions seriously and soon. However, I think we can safely say that if we allowed temperatures to rise 10°C, then the island's economy will not grow forty-four times larger over the next two centuries. But let's set aside Weitzman's theoretical 10°C nightmare scenario. Even unnervingly realistic possibilities, such as an increase of 3°C or 4°C, would almost surely cripple economic growth in Puerto Rico—and nearly everywhere else around the world.

Most IAMs assume that economic growth will proceed in the future much as it has in the past. But numerous scientists and climate economists found it hard to believe that global warming wouldn't stifle growth and elevate the SCC, and their subsequent research vindicated their skepticism. The issue burst into full view in 2015 with the publication of two influential studies.

First came a paper from Stanford researchers Frances Moore and Delavane Diaz. In a press release from the university, Moore said, "For 20 years now, the models have assumed that climate change can't affect the basic growth rate of the economy. But a number of new studies suggest this may not be true. If climate change affects not only a country's economic output but also its growth, then that has a permanent effect that accumulates over time, leading to a much higher social cost of carbon." Much higher indeed. While acknowledging uncertainties and the wide range of possible SCCs, Moore and Diaz estimated that the social cost of carbon should have been $220, not the $37 SCC being used by the U.S. government at the time.

On the heels of Moore and Diaz came a blockbuster study from Stanford's Marshall Burke and the University of California, Berkeley's Solomon Hsiang and Edward Miguel. They crunched economic and climate numbers from 166 nations over the fifty years between 1960 and 2010. Painstakingly the trio carved away extraneous factors until what remained was a record of how each country's economy performed over the years as temperatures rose and fell—but mostly rose, thanks to climate change.

The authors limit their data to a handful of the direct and relatively simple effects of higher temperatures, such as changes in agricultural output, labor productivity, and health attributable to heat. The study finds that a few colder nations will experience net benefits for a while, until their temperatures cross the optimal annual average temperature threshold of about 13°C (55°F), "where economic performance peaks," according to the authors. "Then warming above this temperature causes economic productivity to decline with a rate that accelerates the hotter and hotter a country gets," as the authors put it in an FAQ that followed the study. But most countries will experience net costs soon, if they haven't already.

The authors note that it's crucial to consider what they left out of their study: the more complicated but costly consequences of warming, such as sea level rise, intensified storms, and loss of ecosystem services. These omissions mean that the study significantly underestimates the degree to which warming will diminish growth and how high the SCC should be. However, just because the results are surely underestimates doesn't mean they aren't startling.

"On net, we project that the global economy will do much worse because of climate change, with global average incomes 23 percent lower in 2100 with climate change relative to without it." There's the headline from Burke, Hsiang, and Miguel; the 23 percent figure showed up in literal headlines in many media outlets. The authors base this result on a scenario in which average global temperatures rise to 4°C above preindustrial levels by 2100. Given the authors' conclusion that the decline of growth rates accelerates as the climate gets hotter, that 23 percent number would probably accelerate upwards if global temperatures exceed 4°C.

So what do Burke, Hsiang, and Miguel have to say about the conventional wisdom that unstoppable economic growth will overcome all obstacles? Their answer lies in the reply they give in the FAQ to the following question: "If we grow slowly one year, don't we usually catch up soon after?" The authors respond: "[We] find that reduced productivity today (due to higher temperatures) translates into reduced production in later years. In other words, we find that economies do not 'catch up.' This result is consistent with earlier work looking at temperature and growth." Also recognizing the vulnerability of growth rates is Nicholas Stern, he of the famed Stern Review and one of the world's leading climate economists. He points out the weighty implications for policy: "Clearly, growth itself can be derailed by climate change from business-as-usual emissions. So the business-as-usual baseline, against which costs of action are measured, conveys a profoundly misleading message to policymakers that there is an alternative option in which fossil fuels are consumed in ever greater quantities without any negative consequences to growth itself."

Burke, Hsiang, and Miguel drop another numerical bombshell that did not make the headlines. This finding could so drastically raise the SCC that I'm going to quote one relevant passage in full. And remember that even these eye-popping numbers are an underestimate due to the indirect consequences of warming that the authors omitted from their study.

> Our results suggest the economic cost of climate change is much larger than previously thought. There are several models that are used to compute the economic costs of climate change and we directly compare our results to the three leading models used to set U.S. regulations (DICE, FUND, and PAGE). We find that our results suggest economic losses roughly 2.5–100 times larger than these widely used models. Differences are particularly stark at lower levels of warming. For instance, we show that substantial losses can occur even for end-of-century temperature increases of less than 2C.

Losses 2.5 to 100 times larger. Taking discounting into account and using a middle-of-the-range approach, that means the SCC should be between $200 and $400 instead of around $40, the last federal SCC before the Trump administration gelded the social cost of carbon and deep-sixed the interagency working group. If IAMs incorporated such colossal damage estimates, the models would spit out numbers that would overcome even unconscionably high discount rates and absurdly optimistic projections of future wealth. The stunningly higher SCC would veritably shout at us to start reducing emissions as if our economy depends on it.

THE "WHEN TO STOP RULE"

Does that family on the five-acre lot want to chop down a couple of pines for firewood? Sure, no problem. Does that little software design company want to toss some old computers containing heavy metals into the landfill? Sure, no problem. Does the neighborhood grocery store want to expand by bulldozing that adjacent seasonal wetland? Sure, no problem. Do all eight billion of us humans and all the world's businesses want to do all these sorts of things? Now we've got a problem.

Scale matters. If late one night a single inebriated frat boy wobbles over to the Hudson River and pees into it, the old adage "dilution is the solution to pollution" would hold true. But if everyone in New York City used the Hudson as their toilet, no adage could induce me to swim in that river. Scale is the first principle of sustainability economics.

Orthodox economics also acknowledges the decisive role of scale but only at the micro level. Standard theory recognizes that producers can reach a point at which the extra profit from making one more widget is not worth the extra cost of making it. This constitutes a core concept of neoclassical microeconomics: the "when to stop rule." This rule tells us that there is a point at which costs and benefits balance and we get optimal satisfaction, a point at which more becomes unproductive.

However, as Herman Daly and Josh Farley point out in their book, *Ecological Economics*, "In macroeconomics, curiously, there is no 'when to stop rule,' nor any concept of the optimal scale of the macroeconomy. The default rule is 'grow forever.'" But can we grow forever on a finite planet without reaching a point at which the costs of more outweigh the benefits? Their answer is "no," and they think in some regards we've already reached the "when to stop" point.

And there it is. We've come face-to-face with the dreaded "limits to growth" issue, the controversy that makes the principle of scale a subject of acrimonious debate.

Before we get to the acrimony, let's talk terms. Definitions of growth sprawl all over the place and include both lay and scholarly formulations. The standard version goes something like this: economic growth is the increase over time in the dollar value of market goods and services. Almost everyone measures growth as the percentage rate at which the inflation-adjusted GDP is increasing. Almost universally, societies pursue this kind of growth, and most do so with the zeal of a dog chasing a squirrel. But from a sustainability economics point of view, this definition harbors dangers.

One is the reliance on GDP, which does a poor job of measuring economic health and social welfare. Another hazard lurks in the focus on market value, which can shortchange nonmarket goods. In theory, an economy could avoid such pitfalls and be sustainable while adhering to the conventional growth definition. In reality this never happens. Still, some people and businesses are working to achieve sustainable conventional growth, or what some call "green growth," definitions of which vary widely.

The key to green growth is "decoupling." Decoupling basically means growing while using fewer material resources and producing less waste. (Remember, greenhouse gases are waste.) The ultimate goal is to decouple to the point that the economy can grow indefinitely but sustainably. Clearly, we should decouple early and often, but many green growth enthusiasts seem to think that all it will take is a bit of recycling here and some wind turbines there—no fundamental changes to the market required. I'll call this viewpoint "green growth lite." Its champions conjure an economy that doesn't break stride in its expansionary sprint, largely propelled by the desultory adoption of some green tech and lacking in systemic change.

Green growth lite boosters claim we're well on our way to a decoupling nirvana, noting that in recent years several developed nations have continued to grow their GDP while reducing their resource use and waste. But this optimism overlooks many issues, most obviously the fact that those vaunted reductions have been modest and scattered. And even those modest reductions largely evaporate when you take into account wealthy nations' predilection for exporting externalities to developing nations; the low-income countries do the dirty work while high-income countries consume the products manufactured beneath the foreign smokestacks.

A study by economists Enno Schröder and Servaas Storm that accounted for this offshoring of greenhouse gases found no evidence of the decoupling of global GDP and global carbon emissions between 1995 and 2011. As Schröder and Storm write, "Without a concerted (global) policy shift to deep decarbonization, a rapid transition to renewable energy sources, structural change in production, consumption, and transportation, and a transformation of finance, the decoupling will not even come close to what is needed." They also write, "The key insight is that marginal, incremental improvements in energy and carbon efficiency cannot do the job and that what is needed is a structural transformation . . . Radical change within a limited time span is what we need, and this needs collective action and a strong directional thrust which 'markets' or 'private agents' alone are unable to provide." At the very least, we should follow the precautionary principle and not count too much on decoupling unless and until it proves it is up to the job.

Schröder and Storm's biting takedown of green growth lite enjoys considerable corroboration, and not just in the context of climate change. Take the UN report on biodiversity that made waves in 2019. The authors conclude that anything short of "transformative change" will not stanch the hemorrhaging of species. First among the key leverage points listed by the researchers is "enabling visions of a good quality of life that do not entail ever-increasing material consumption." And the last point made in the report's summary is, "A key constituent of sustainable pathways is the evolution of global financial and economic systems to build a global sustainable economy, steering away from the current limited paradigm of economic growth."

So the term "green growth" can refer to anything from green growth lite to a far more transformational version that sounds a lot like sustainability economics. This variability renders the term too vague to help much with our understanding of growth. The standard definition of economic growth is more settled, but whether or how much sustainability can be attained by using that as a guide remains debatable. So in the context of this discussion I'll use homegrown definitions. "Growth" refers to quantitative expansion of the economy based on an unsustainable use of materials and energy. "Development" refers to qualitative improvement of the economy based on the sustainable use of materials and energy. Broadly speaking, neoclassical economics leans toward growth, and sustainability economics favors development.

Questioning endless growth alienates nearly everyone. Republicans, Democrats, libertarians, socialists, Tea Partiers, progressives, plutocrats, laborers—they all clamor for more growth, though often for different reasons. Some allow that we should take steps to make growth more environmentally sustainable, but usually that goal comes in a distant second. People are far more focused on the anticipated benefits of growth; it will create jobs, swell profits, provide material abundance, reduce inequality, secure a prosperous economy for the indefinite future, lift people out of poverty, foster world peace, and maybe even turn lead into gold. Standard economics ignores or dismisses the notion that there are biophysical limits to growth that will interfere with reaching these goals.

Sustainability economics begs to differ. Its proponents believe that limits do exist and therefore growth can't achieve all the fine goals listed above. But don't despair. Development can help us reach those goals, except the one about turning lead into gold. In fact, in the long run, shifting to development is the only way we can reach those goals. Still, it won't be easy and it won't happen overnight. Our economy and our culture have a lot of evolving to do.

But while we're evolving, what happens to our fellow humans who don't have enough to eat? What about sick people who are dying because they lack decent health care? What about children who can't go to school because they have to work to help support their families? How can we deny these people a minimally decent life even if it's powered by coal or involves destroying a rain forest? Shouldn't these people reap the benefits of growth, just as wealthy societies did in the past? Isn't it a moral imperative?

Such questions quickly crop up during discussions of limits to growth. Sometimes they're voiced by the poor themselves or by sincere advocates for the poor. At other times they're voiced by the lords of growth who use alleviation of poverty as a smoke screen to mask their self-serving behavior. But regardless of the source, these are valid and pressing questions that demand answers.

So, yes, absolutely yes, alleviating poverty is a moral imperative. But so is reducing environmental harms, which are greatly increasing poverty; if left unchecked, climate change will likely become the world's greatest driver of poverty. The partial answer to this dilemma is to accept the need to burn some more coal and destroy some more rain forest but to keep such unsustainable growth to a minimum. It should increasingly be reserved only to help the poor escape their penury. Currently many of the people who benefit the most from the burning of coal and the destruction of rain forest are reasonably well-off, if not downright wealthy.

More important, as we wind down growth, we must ratchet up development. Rich nations must contribute most of the funding for this transition, both because they have most of the resources and because they are responsible for most of the environmental damage that necessitates the transition. This includes rapidly converting their own economies as well as providing sufficient aid to poorer countries so they can wean themselves off growth. If rich nations provide enough aid, impoverished nations can skip some of the coal-and-rain-forest-destruction growth stage and leapfrog to development.

The need for wealthy countries to step up has long been recognized and has been enshrined in one international accord after another, but the financial commitments have been inadequate and the actual delivery of the money has been truly sorry. Back in 2009 the developed nations agreed to drum up $100 billion a year by 2020—a modest-enough amount considering the gravity and scale of climate change—but nowhere near that much has been collected and even less has made it to actual projects. The numbers are a little fuzzy, but it seems that only a few billion dollars a year have reached the developing nations. At 2019's key international climate meeting these increasingly desperate countries pushed hard for assistance, but once again, the wealthy nations blocked progress.

The rich countries of the world clearly can afford to bankroll sustainable development in low-income nations. America spends between $750 billion and $1 trillion a year on our military alone—the same military that has declared climate change to be an urgent national security issue—yet the Trump administration has declared that it won't provide any more American dollars to the international effort to help poor countries reduce emissions and adapt to climate change. Note that Aramco, the Saudi Arabian oil company, enjoyed a net income of about $111 billion in 2018; it seems only fair that oil producers contribute to climate action in the developing world. If the failure to sufficiently fund climate work forces too many struggling nations to opt for growth instead of development, the world may end up with an outbreak of retrograde infrastructure that will entrain growth for decades to come and massively increase emissions.

Case in point: China's Belt and Road Initiative (BRI), which could end up being the largest infrastructure project ever. Mainly aimed at nations that desperately need infrastructure, the plan aims to finance and build seaports, mines, bridges, railroads, airports, and the like throughout Asia and eventually all around the world. According to the Asian Development Bank, infrastructure needs in Asia alone will run about $26 trillion over the next decade. Chinese officials tout the BRI as a key to lifting people out of poverty and stimulating global trade. Other observers see it as a power grab by China and an engine of corruption in weakly regulated nations.

China's president, Xi Jinping, declared that the BRI will be "green" and "healthy," but so far reality declares otherwise, as evidenced by the hundreds of power plants spawned by the initiative. Numerous studies of BRI power plants reveal that so far a large majority use fossil fuels, mostly oil and coal. Typically such plants operate for a long time—coal facilities have an average life span of about forty years—so BRI is currently on a trajectory to lock in decades of profuse greenhouse gas emissions in the quest for growth. As climate economist Nicholas Stern said in 2018, "The more than 70 countries that are signed up to the Belt and Road Initiative have an average GDP of around one-third that of China. If they adopt China's development model, which resulted in a doubling of China's GHG [greenhouse gas] emissions in the first decade of the century, it would make the emissions targets in the Paris Agreement impossible." To secure a livable climate, wealthy nations must stave off BRI-like growth by amply supporting development in low-income countries.

The idea of slowing economic growth and even easing it to a stop has been around almost since the beginning of the Industrial Revolution, though the concept has always been relegated to the margins of our growth-hungry society. A famed proponent in the nineteenth century was the political economist John Stuart Mill, who dubbed his vision the "stationary state." (Today's heirs to Mill's vision typically use the term "steady-state economy.") He espoused a stable but not static economy, one that would continually evolve but remain indefinitely in a state of dynamic equilibrium rather than growing. As he wrote in 1848, "It is scarcely necessary to remark that a stationary condition of capital and population implies no stationary state of human improvement. There would be as much scope as ever for all kinds of mental culture, and moral and social progress; as much room for improving the Art of Living and much more likelihood of its being improved, when minds cease to be engrossed by the art of getting on."

Most of Mill's contemporaries dismissed the stationary state as a baleful notion that would halt progress, but perhaps they made the same mistake that many of today's critics of a steady-state economy make: picturing a steady-state economy as simply a growth-dependent economy that suddenly stops growing. True, that's not a pretty picture, but that's also not an accurate portrayal of a steady-state economy. That's like imagining a new transportation plan based on walking, bicycling, and transit imposed onto an existing infrastructure and sprawling layout that were designed for cars. Such a mismatch would produce a huge muddle. Just as a transportation system meant for walking, transit, and biking would require different infrastructure and a different layout, so would a steady-state economy require many new institutions, new ways of doing business, and new attitudes toward consumerism.

To minimize disruption, steady-staters think we should make such fundamental changes as gradually as possible, but time is growing short. Even back in the nineteenth century Mill worried that society would wait too long and end up facing the overgrown era we have now entered, an era in which we have no sane choice but to briskly move from growth to development. As he wrote, "I sincerely hope, for the sake of posterity, that [humanity] will be content to be stationary, long before necessity compels them to it." For a more current comment, let's hear from Greta Thunberg, the renowned teenage climate activist from Sweden. In 2019, in her speech to the United Nations, she told the delegates, "People are dying. Entire ecosystems are collapsing. And all you can talk about is money and fairy tales of eternal economic growth."

Though still a feisty minority, a growing number of people recognize the need to put the brakes on growth and nurture development, which would entail transitioning to a steady-state economy or at least greatly slowing growth. They brim with ideas about how to accomplish this shift. For a sampler, let's turn to Gus Speth, an eminent figure in the development camp, who served as a White House adviser under Jimmy Carter, dean of the Yale School of Forestry and Environmental Studies, and administrator of the UN Development Programme. Here are just a few of his suggestions, all taken from just one paragraph in just one of his books: redesign corporations, bolster public services, institute fees and caps on the extraction of virgin materials, support organized labor, dial down the stock market's obsession with maximizing shareholder returns, provide a sufficient minimum wage, expand parental leave, infuse trade agreements with strong environmental and social provisions, and restrict advertising.

Currently Speth co-chairs the Next System Project, which seeks to develop systemic solutions to our systemic problems. Some of their efforts aim at the connections between growth and climate change. Take their push for what they call "energy democracy," which involves shifting ownership of utilities from private investors to the public in the form of local governments or local co-ops. That would relieve the pressure to expand in order to inflate returns for distant shareholders, giving the local utility the freedom to simply do what's best for the community, from lowering rates to bringing on more renewables to filling utility jobs with local workers. Energy democracy is not just a concept, either; many local public utilities are already operating successfully.

With no shortage of workable ideas and a decent start at implementation, the goal of curbing growth in ways that create a net gain for society seems attainable. However, this optimistic side of limits to growth usually gets drowned out by the chorus of critics who think the end of growth means humanity will revert to stone tools and slink back into caves. As noted above, the rejection of limits to growth has come from all political and philosophical quarters, but the attacks by half of the groups mentioned above—Republicans, libertarians, Tea Partiers, and plutocrats—have been particularly ferocious. After all, limits strike at the heart of their beloved vision of an infallible market economy.

The vitriol hurled at the much-discussed 2018 scholarly paper "A Good Life for All Within Planetary Boundaries" provides a representative case. The authors plumb "the challenge of how to achieve a high quality of life for over 7 billion people without destabilizing critical planetary processes," as they put it. They set thresholds for major biophysical features, such as freshwater supply, climate change, and biological diversity, which humanity should not exceed. They also set thresholds for social features, such as nutrition, income, and education, which would provide people with a decent but not extravagant material life and a high-quality personal and community life. Then the authors plow through data from more than 150 nations to see how well each is doing. The results are sobering. "We find that no country meets basic needs for its citizens at a globally sustainable level of resource use." And growth is not the solution; it is the main problem.

Now, about that vitriol. Wesley Smith's opinion piece in the venerable conservative magazine *National Review* offers a representative example. In it he assails "A Good Life for All Within Planetary Boundaries." Smith's title, "Environmentalists Push Global Wealth Redistribution," immediately springs one of free marketeers' most feared bogeymen on his readers. He follows the title with this opening line: "The environmental movement wants to make the rich West much poorer so that the destitute can become richer." Elsewhere in the article Smith continues to pound the table about the authors' redistributionist heresies, yet redistribution is hardly the point of their paper, which concerns growth. I think Smith's hair-trigger leap to redistribution reveals that one reason many of the world's affluent promise growth will lift all boats is that this false promise allows them to avoid the fraught topic of inequality.

Smith does eventually turn his guns on the study's main subject. Here are a few lines from Smith's piece that convey the flavor of his commentary. "This means limits, limits, limits!" "The goal clearly is a technocracy that will undermine freedom, constrain opportunity, not truly benefit the poor, and materially harm societies that have moved beyond the struggle for survival." "In other words, growth is out. We must live within economic and social systems strictly limited by arbitrary boundaries on the use of resources established by 'the experts.'"

Setting aside Smith's hostility, I would like to focus instead on something in that last sentence of his, a loaded word whose implications get lost in his hyperventilation: "arbitrary."

This word and the context in which Smith uses it, particularly his sneering quotation marks around "the experts," suggests contempt for scientists and the science that indicates humankind is violating biophysical boundaries. But the boundaries discussed in the paper are not arbitrary. Nor are those described in many other studies about the limits to growth. Yes, uncertainty abounds, à la climate science, and a reasoned questioning of the results is a welcome part of the process. But the results are empirically based, not arbitrary.

"A Good Life for All Within Planetary Boundaries" draws on the influential 2009 article "Planetary Boundaries: Exploring the Safe Operating Space for Humanity," written by a stellar international assemblage of twenty-eight scientists. The 2018 study also incorporates prolific ongoing research regarding planetary boundaries, a project hosted by the Stockholm Resilience Centre. The 2009 article and the project it spawned identify nine major boundaries—the biophysical thresholds referred to in the 2018 study. "Crossing these boundaries could generate abrupt or irreversible environmental changes," according to project scientists. The latest project update calculates that runaway growth has already pushed the planet past four of those nine thresholds.

"Human beings and the natural world are on a collision course." This sounds like another quote from the planetary boundaries research, but it's actually the first line in another influential article: "World Scientists' Warning to Humanity." You can imagine what the rest of the warning said; the word "finite" came up a lot. The Union of Concerned Scientists published the scientists' warning in 1992. More than 1,700 scientists from seventy-one countries signed the document, including 104 Nobel laureates—the majority of then-living winners of the Nobel Prizes in the sciences.

Fast-forward twenty-five years. Late in 2017 a group of international researchers published "World Scientists' Warning to Humanity: A Second Notice." The authors had followed up on many of the worrisome environmental issues cited in the first scientists' warning to see how humanity was doing. Not well, I'm afraid, not well at all. As the researchers write, "Since 1992, with the exception of stabilizing the stratospheric ozone layer, humanity has failed to make sufficient progress in generally solving these foreseen environmental challenges, and alarmingly, most of them are getting far worse. Especially troubling is the current trajectory of potentially catastrophic climate change." In a follow-up column discussing the second notice, the paper's authors write that "transformative change is essential, whereby humanity abandons the pursuit of economic growth as the overarching guide to public policy. We need a new development paradigm to ensure that economies deliver well-being while respecting both social and planetary boundaries." Seeking endorsements, the authors sent out one tweet and directly contacted about forty scientists. Within forty-eight hours some 1,200 scientists had added their voices to the second notice. And the endorsements have kept pouring in. As of 2020, more than 23,000 scientists from more than 180 nations had signed on to show their support of the second notice.

Anticipating the usual attacks from the Wesley Smiths of the world, the lead author of the second notice, William Ripple, writes, "Some people might be tempted to dismiss this evidence and think we are just being alarmist. Scientists are in the business of analyzing data and looking at the long-term consequences. Those who signed this second warning aren't just raising a false alarm. They are acknowledging the obvious signs that we are heading down an unsustainable path."

Reasons for our self-destructive behavior range from innocent ignorance to knowing greed, but in the next chapter I want to focus on just one. It's an essential economic concept that enables unsustainable growth.

10

ORBITING GIANT MIRRORS

Whale oil provided the lighting to read the break-through novel of 1870, the story of Captain Nemo in *20,000 Leagues Under the Sea*. That was also the year of the foundation of Standard Oil. The result of that foundation is that we didn't hunt the whales to extinction, but instead turned to kerosene to light the latter part of the 19th century, moving to electricity only in the 20th.

It really isn't hyperbole to insist that John D. Rockefeller saved the whales by his making mineral oil products so much cheaper than the cetacean-derived equivalent. And that's really all you need to know to understand Earth Day and what to do about it.

We need to be as viciously capitalist and free market as we can to save the planet.

The above passage comes from an Earth Day 2018 article written by Tim Worstall, a climate-change-won't-be-that-bad advocate and senior fellow at the conservative Adam Smith Institute. I'm going to skip over Worstall's provocative choice of a founding father of big oil as an environmental hero—after all, oil may have saved some whales, but it has also caused a wee bit of environmental damage—and get to the point: substitutability. That is the term for the economic process illustrated by Worstall's tale about the replacement of whale oil by petroleum. As we'll see, this concept also includes less direct substitutions, such as money replacing a natural resource like whale oil. Substitutability is a load-bearing concept supporting the belief that our economy can grow forever.

> Greater consumption due to increase in population and growth of income heightens scarcity and induces price run-ups. A higher price represents an opportunity that leads inventors and businesspeople to seek new ways to satisfy the shortages. Some fail, at cost to themselves. A few succeed, and the final result is that we end up better off than if the original shortage problems had never arisen.

This sunny summary of how substitutability theoretically operates comes from Julian Simon, who was probably the most famous and undeniably one of the most fervent of the so-called Cornucopians. A longtime business professor and a senior fellow at the Cato Institute, Simon saw the world as a cornucopia of endless bounty. More people, more growth, more consumption—it's all good, and there are no limits as long as we let the market work its substitutability miracles. But sustainability economists feel less assured about substitutability. Though they think it must play an important role in achieving a sustainable economy, they recognize its many deficiencies and emphatically don't think it will enable endless growth.

A few years ago I spent some time with commercial fishers in the Copper River region of Alaska, famous for its sockeye salmon. I was researching the system they use to limit their fishing in order to maintain a profitable sockeye fishery for the foreseeable future. But let's imagine that the Copper River fishers convert to Cornucopianism, renounce the heresy of limits, and proceed to overfish Copper River sockeye to the point the species becomes scarce. Following the rules of supply and demand, as sockeye numbers go down, their price would go up. Eventually the ballooning price would drive away so many customers that catching sockeye would become a losing proposition. This is the market's way of telling the fishers that they had better find a substitute.

That substitute could take many forms. Maybe fishers would turn to pink salmon, at least until limitless fishing in turn drove pink numbers so low and their price so high that the fishers had to find yet another substitute. Maybe coho salmon would be their next target, and then chum salmon. After running out of salmon species, the fishers might switch to non-salmon fishes. And once there weren't enough desirable fish of any sort left, the fishers would try to find other careers and consumers would try to find other sources of protein. Simon says this chain of events represents the market providing society with new opportunities for improvement. I say such a chain of events often does more harm than good.

Substitutability's most glaring defect is that it offers little incentive for the private sector to produce substitutes for many nonmarket goods, notably ecosystem services. (If substitutes even exist. We'll get to that.) Take a stable climate. Though the global supply of climate stability has been shrinking rapidly, the market has not responded with the necessary magnitude of demand for substitutes. Even after the scale of the climate crisis became obvious, far more investment has gone into fossil fuel production than into energy efficiency and clean energy development. And many of the dollars that did manage to find their way to climate-friendly efforts have either come from government or have been prompted by government regulations or the prospect of regulations. The increasing scarcity of climate stability hasn't resulted in the price run-ups Simon cites. How could it when there is no one paying for a stable climate in the first place? Without price run-ups the process of substitutability never gets started.

Even when we are dealing with market goods, substitutes often fail as one-to-one replacements, as our salmon situation illustrates. For example, most consumers would consider a switch from sockeye to chum salmon a downgrade. Often called "dog" salmon, chum salmon officially get their nickname from the prominent canine teeth the males develop during spawning, but the nickname is also apt because chum is the least desirable salmon species, something Alaska Natives traditionally fed to their sled dogs. Substituting chum salmon for sockeye would not make consumers better off.

Substitution also stumbles at times because real consumers don't always turn to substitutes the way Econ 101 says they will. As we saw in chapter three, overfishing of bluefin tuna is pushing that coveted species toward commercial extinction, with populations having shrunk to a small fraction of natural levels. As Econ 101 predicts, the price has risen in response to the bluefin's rarity. But Econ 101 did not predict that some consumers would continue to demand primo wild tuna seemingly no matter how high the price goes; you may recall that bluefin has sold for as much as $4,900 per pound. Substitution has not ridden to the rescue. With bluefin numbers so low, only a few wealthy people get to eat bluefin, and there aren't enough of these top predators to fill the species' ecological niche. Again, we're not better off.

One of substitutability's weakest soft spots has likely already occurred to you: What about essential resources for which there is no reasonable substitute? Fresh water, for instance, given that we can't just drink dirt or drill for Gatorade to quench our thirst once water gets scarce and expensive. Just look at what's going on in America's Southwest, where long-standing water problems are being worsened by global warming. Despite decades of trying to squeeze ever more water out of the already overdrawn Colorado River and the region's more recent push to conserve water, the people of the South-west are in deep trouble.

Some scientists think the region has entered mega-drought territory. A mega-drought produces conditions similar to the worst droughts of the twentieth century, but instead of easing after a brief spell, the hot and dry conditions roast the land-scape for several decades. The two researchers who coined the term set the bar at twenty years; any severe drought per-sisting for at least twenty years qualifies as a mega-drought. The current Southwest drought began in 2000. "This will be worse than anything seen during the last 2,000 years," writes Toby Ault, a Cornell University earth science profes-sor and a coauthor of a study that predicted a mega-drought for the Southwest. The possible consequences range from destructive to unbearable. In the unbearable scenario, farm-ing becomes untenable, dust storms darken the skies, most plants perish, and inhabitants leave en masse. As University of Michigan climate scientist Jonathan Overpeck says, "If we don't deal with climate change, the Southwest will become a place of exodus."

Sadly, the Southwest's predicament is far from unique. Reports from around the world about small towns, large cit-ies, and entire regions running short on water are cropping up with increasing frequency as the impacts of global warming mount. A 2018 paper from the World Bank and the UN notes that 70 percent of the world's people already suffer from water scarcity at least once a month every year. And no substitute exists. We can't proclaim, "Let them drink cake."

IN TECH WE TRUST

The increasing scarcity of water often leads to thoughts of desalination. Unfortunately, for now at least, turning sea water into fresh water costs too much to offer a widespread answer to our water woes, but it does bring up the concept of technological substitution, a pillar of substitutability along with the natural substitution we've been discussing. Many believers in endless growth have an unwavering faith that the market will stimulate technological innovation and provide whatever substitutes we may need. In his influential 1996 book, *Ultimate Resource 2*, Julian Simon explains that we could feed an ever-growing population by using new technologies, such as "orbiting giant mirrors that would reflect sunlight onto the night side of the Earth and thereby increase growing time."

Yes, he said "orbiting giant mirrors." Yes, this seems a bit wacky. But, to be fair, I cherry-picked one of Simon's more outlandish notions. I mention the example of the mirrors because it conveys the credulity of many Cornucopians when it comes to the capacity of technology to infallibly supply us with substitutes. Often this confidence isn't attached to any specific invention. Rather, it's a fuzzy, blanket presumption that technological ingenuity will take care of things. Consider a couple of excerpts from Simon's writings in the 1990s: "Technology exists now to produce virtually inexhaustible quantities of just about all the products made by nature." And "We have in our hands now—actually in our libraries—the technology to feed, clothe, and supply energy to an ever growing population for the next 7 billion years." The latter quote became notorious, perhaps because seven billion years seemed a bit hyperbolic, what with the sun due to fizzle out before that. Evidence emerged later that Simon had probably made a mistake and had meant to write "7 million years." Well, that's a start, but he still would have needed to remove a lot more zeroes in order to avoid hyperbole.

Incidentally, a fresh take on orbiting mirrors surfaced in recent years, most notably when Democratic presidential candidate Andrew Yang mentioned them during his campaign in 2019. But his version of "space mirrors," as he referred to them, would serve a different purpose than Simon's ag-boosting mirrors. Theoretically, they would reflect some sunlight away from the earth and reduce global warming. But Yang made it clear that they are not ready for deployment and that any such geoengineering measures should only be used as a last resort. Space mirrors are decidedly not central to his climate plan, which emphasizes less literally far-out ideas, such as establishing zero-emission standards for cars, passing a carbon tax, and ending all leases for fossil fuel production on federal lands.

Simon possessed a gift for exuberant expressions of confidence in technology, but more pedestrian declarations of techno-optimism from prominent and powerful people have been cropping up for ages. Case in point, the following scene at one of then House Speaker Paul Ryan's weekly briefings. "The earth is warming up. What should we do?" The question for Ryan came from the press section, but the soft, high-pitched voice didn't sound like that of a reporter. In fact, the query came from the seven-year-old daughter of one of the reporters; it was Bring Your Child to Work Day. "I think technology is the best answer to this question," replied Ryan. Easy. Next question.

The flow of techno-optimism from Ryan's fellow Republicans started swelling into a flood late in 2018, when Americans' concern about global warming began escalating. The public increasingly perceived climate denial as troglodytic, and surveys revealed that even many young Republicans were warming to the reality of climate change. Fearful of electoral backlash, some members of the GOP, including a few long-time opponents of climate action, grudgingly allowed that maybe temperatures were rising, though, some said, not necessarily from human activity, and perhaps something should be done.

And what is that something? Technological innovation, of course, driven by the market. And certainly not government regulation. In fact, much of the Republicans' newfound passion for climate tech seems like a preemptive strike designed to quell any regulatory notions that people might have, especially those associated with the much-reviled Green New Deal. Consider the headline atop a December 18, 2018, *New York Times* op-ed written by U.S. senator John Barrasso, the Republican who chairs the Environment and Public Works Committee: "Cut Carbon Through Innovation, Not Regulation." Or how about in 2019, when Republican senator Lamar Alexander stood on the floor of the U.S. Senate and proclaimed that "the United States should launch a new Manhattan Project for clean energy"? Or take Republican Matt Gaetz's industry-friendly riposte to the Green New Deal: his "Green Real Deal." A staunch Trump ally and the sponsor of a 2017 bill calling for the elimination of the EPA, Gaetz told Bloomberg News in 2019, "Unilaterally disarming the American economy through crushing regulations will empower Washington but few others. Our rise in global leadership on climate must be fueled by American innovators."

I certainly don't mean to discount innovative tech. Technological substitutes are vital to the effort to become more sustainable. Replacing coal with solar and incandescent light bulbs with LEDs would help enormously. But innovative tech alone won't be enough, and I do decry the politicians who invoke it as a silver bullet that precludes the need for other tools, such as regulations and international agreements.

For one thing, technological substitutes share many of the same failings as natural substitutes, with perhaps the most insidious being unintended consequences. Palm oil plantations in Indonesia provide a distressing example. For the details of this tragic tale I'm drawing on a compelling 2018 article by Abrahm Lustgarten, "Fuel to the Fire," that was a collaboration between ProPublica and *The New York Times Magazine*. Lustgarten chronicles the unintended consequences of a 2007 law that compelled American gasoline and diesel producers to add a lot more biofuel to their fuel blends. (Biofuels are fuels derived from living matter, usually vegetation, and are officially considered clean and renewable, though many people dispute that official label.) The Bush administration supported the legislation in order to help America achieve energy independence and to score points with the farm lobby. The Democratic Congress supported it for the same reasons and additionally to help mitigate climate change by replacing some petroleum with biofuels.

When Bush and Congress pictured biofuels, they envisioned good old American corn and soy as the fuel stock, but it turns out the Midwest didn't have much unused farmland lying around that could be planted with more corn and soy. Soon American biofuel producers began importing ingredients, notably palm oil from Indonesia.

The whole idea of biofuels becomes murky due to complications, such as the trade-offs involved when you grow fuel stocks on land formerly used to produce food. But one problem with using Indonesian palm oil is stark: multinational corporations and local outfits alike began shearing lush Indonesian rain forests to make room for oil palm plantations. Besides the huge damage to biodiversity and indigenous peoples, the logging released vast amounts of carbon dioxide that had been stored in the forests. But the money was good and the shearing accelerated. Soon the forests were falling at the rate of three acres a minute. By now millions upon millions of acres have been cleared.

It gets worse. Some of the vanishing forest is peatland. Peatlands sequester extraordinary amounts of carbon, about twelve times more per acre than ordinary tropical forest—and ordinary tropical forests are renowned for their ability to store carbon. Disturbing peatlands releases colossal volumes of carbon into the atmosphere.

A number of studies indicate that the use of palm oil in biofuels leads to a net increase in carbon emissions, perhaps a large increase. Former U.S. representative Henry Waxman (D-CA), a longtime congressional supporter of environmental causes, originally approved of the biofuels legislation, but he has come to regret that early approval. As quoted by Lustgarten, Waxman recently said, "We've created a situation that is so contrary to what we had hoped for. We're doing more harm to the environment. It was a mistake."

The confluence of technological substitutes, climate change, and unintended consequences inevitably leads us to the subject of geoengineering, which some consider the ultimate technological solution to global warming. The UK's Royal Society provides this concise definition of geoengineering: "deliberate large-scale manipulation of the planetary environment." So . . . manipulating the planetary environment. On a large scale. Hard to imagine unintended consequences.

One frequently cited example of geoengineering involves sprinkling the oceans with iron dust to stimulate massive blooms of carbon-munching phytoplankton. Or we might build machines to suck carbon dioxide out of the air and store it underground. As noted, the ideas even include orbiting giant mirrors, the sequel, in which the mirrors would reflect a bit of sunlight back into space before it can warm the earth's surface. A variation of this concept would deploy a single mirror that takes Simon's adjective "giant" into uncharted territory; one suggestion envisioned a mirror more than twice the size of Texas. Among the potential obstacles noted was the cost estimate, which ranged rather widely, from $800 billion to $400 trillion. That's like a plumber telling you he can unplug your bathtub drain for somewhere between $80 and $40,000. Not surprisingly, the Intergovernmental Panel on Climate Change labeled such ideas as "speculative, uncosted and with potential unknown side-effects."

I should add that the other two ideas mentioned above are more plausible, as are some other geoengineering approaches. More plausible, but not ready for prime time. As the Royal Society noted in its report on geoengineering: "The safest and most predictable method of moderating climate change is to take early and effective action to reduce emissions of greenhouse gases. No geoengineering method can provide an easy or readily acceptable alternative solution to the problem of climate change." We definitely should explore "deliberate large-scale manipulation of the planetary environment" because we'll need every instrument we can get our hands on. Maybe we'll get lucky and someone will invent the aforementioned atmospheric Roomba. But we should tread carefully and not bank too much on geoengineering, as some current climate plans do. And whenever the idea of geoengineering crops up—especially if it involves orbiting giant mirrors—a discussion of unintended consequences should not be far behind.

NATURE IS COMPLICATED

Tears welled up in my eyes and I stopped the video. It is only a minute and eight seconds long, but thirty or forty seconds was all I could take. Watching little Petey trying to eat the food his parents brought him was heartrending when I already knew he eventually starved to death.

Petey was a puffin. Born on Maine's Seal Island, Petey became famous because researchers set up a "puffin cam" inside his snug burrow so scientists, schoolchildren, bird-watchers, and others around the world could watch this fluffy puffin chick grow until he was ready to fly out into the world. But Petey didn't grow as he should have because he couldn't eat the butterfish his parents provided. Puffins don't take bites out of fish; they swallow them whole. But in the video you see that despite all his struggles, Petey can't fit the butterfish into his mouth because the fish are too big. He needed the smaller hake and herring that usually make up the bulk of a Maine puffin chick's diet.

As weeks went by, Petey's parents continued to bring him butterfish. He desperately kept trying to choke them down, but to no avail. Eventually his efforts faded as he weakened from hunger. Finally one summer day Petey quietly died.

Why did Petey's parents keep giving him butterfish? Scientists had previously tracked his parents and knew that they had successfully fed and raised chicks before. Had the parents somehow lost their ability to hunt hake and herring? When researchers surveyed sixty-four other puffin burrows, the mystery deepened; in most of the burrows the scientists found dead chicks and heaps of rotting, uneaten butterfish. The researchers finally solved the mystery when they studied the waters around Seal Island and found far fewer hake and herring than usual. They also discovered the likely reason most of the hake and herring had disappeared: the water was too warm for them. In fact, the water temperatures in the Gulf of Maine that year turned out to be the highest ever recorded, apparently due to climate change. To understand how this fits into the big picture of substitutability we need to start small.

Phytoplankton are the microscopic, plantlike organisms that form the base of the marine food web. The warming waters have made life hard for the Gulf of Maine's phytoplankton, notably diminishing their vital spring bloom. This in turn has diminished the production of zooplankton, the microscopic marine animals that feed most of the fish, lobsters, whales, and other denizens of the sea that humans—and puffins—value so highly. The biomass of the gulf's zooplankton sank to historic lows the summer Petey died, forcing many hake and herring to travel north in search of food.

Many other fish species also are fleeing northward from the Gulf of Maine in search of cooler waters and adequate food. A National Oceanic and Atmospheric Administration study found that the ranges of seventeen fish species in the region are shifting north at the rapid pace of up to five miles a year. Perhaps the puffins can likewise move north, but such adaptations usually are more complex than people, including mainstream economists, realize. Most animals and plants have complicated needs that go well beyond sources of food. For example, puffins can't establish successful breeding colonies on just any old rock. They favor small islands and offshore sea stacks that are free of predators and that feature soil in which they can dig their burrows. These and other requirements aren't always easy to find. No wonder puffins typically return to the same colonies year after year and generation after generation.

Climate change is degrading habitats on land and sea around the globe. In 2018, a massive study looked at the impacts of such habitat impairment on 115,000 species of plants and animals, estimating the extent to which climate change would alter their range by 2100 under various warming scenarios. Consider the case of insects, as reported by *Science Daily*. Here's a quote from the study's leader, Rachel

Warren, of the Tyndall Centre for Climate Change Research at the University of East Anglia.

> Insects are particularly sensitive to climate change. At 2°C warming, 18 per cent of the 31,000 insects we studied are projected to lose more than half their range.
>
> This is reduced to 6 per cent at 1.5°C. But even at 1.5°C, some species lose larger proportions of their range.
>
> The current global warming trajectory, if countries meet their international pledges to reduce CO_2, is around 3°C. In this case, almost 50 per cent of insects would lose half their range.
>
> This is really important because insects are vital to ecosystems and for humans. They pollinate crops and flowers, they provide food for higher-level organisms, they break down detritus, they maintain a balance in ecosystems by eating the leaves of plants, and they help recycle nutrients in the soil.

And thus Professor Warren brings us to the big picture: our ecosystems and the life-giving services they provide.

Petey died for lack of ecosystem services. There was no substitute for the missing hake and herring, and the hake and herring went missing because there was no substitute for the depleted phytoplankton and zooplankton that hake and herring feed on.

I wouldn't count on market innovation to provide viable substitutes to heal what ails ecosystems like the Gulf of Maine. Even if it were possible to devise reasonable substitutes, they would never happen if we relied on the market. It doesn't pick up the signals phytoplankton and zooplankton are sending by their scarcity. These microscopic organisms are not only invisible to the human eye; they are invisible to the invisible hand.

THE PRODUCTION FUNCTION MALFUNCTION

It's time for you to meet the production function. I swore not to include any equations in this book and I'll stick to that vow, for my sake as much as yours. But I feel compelled not to skip this function because in purporting to show how inputs are turned into outputs, the orthodox version of this central economic process reveals ways in which the neoclassical view of substitutability is out of touch with biophysical reality. In introducing the function I draw heavily on the work of ecological economists Herman Daly and Josh Farley.

The function is simple. The quantity of some item that is produced equals the factors that go into its production, traditionally labor, capital, and resources. You can produce more widgets if you add more workers, invest in more equipment, use more resources, or some combination thereof. The function views resources narrowly, as mere material inputs, such as trees for producing lumber or copper for your smartphone.

Sustainability economists find this function simple to a fault. I'll mention just three of the omissions they cite. One, the function overlooks waste, even though almost all production generates some kind of unwanted by-product—such as greenhouse gases. Two, the function leaves out energy resources, which are always required, even if it's only the energy to produce the food that feeds the body that taps the keys on a computer keyboard. Three, the function recognizes only human-made capital and omits natural capital, such as ecosystem services.

Sustainability economists also see the classic production function as too abstract. They think biophysical reality gets lost amid the math, which seems designed more for the convenience of economists than for the purpose of clarifying the way the world works. For example, in the transcendent realm of the function, you can substitute any factor of production—labor, capital, or resources—for another. This conveys a profoundly inaccurate view of the power of substitutability to sustain or grow output.

Let's return to our Copper River salmon fishers and the hypothetical scenario I set up in which they forswear the limits they had established, adopt Cornucopianism, and overfish the local populations of one salmon species after another. According to the logic of the neoclassical production function, the fishers need only substitute one or both of the other factors of production—labor and capital—to compensate for the reduction in resources (i.e., the dwindling fish stocks). For example, a fisher could buy a second boat (capital), hire a second crew (labor), and, abracadabra, she'd be netting plenty of fish again and her catch (output, in the function) would return to its former level or maybe even grow.

Unless there aren't enough fish. The standard production function reflects a preindustrial era when the number of fish in the sea was functionally infinite. But in the twenty-first century, resources often are the limiting factor. Without enough fish, no amount of labor and capital will put salmon on the grill.

The economist Robert Ayres writes, "The fundamental problem is that neoclassical economic theory has no role for physical materials, energy or the laws of thermodynamics." A few lines later he adds, "It is fundamentally a theory about relationships between immaterial abstractions. Moreover, standard theory assumes that scarcity does not exist in reality, because any threat of scarcity is automatically compensated by rising prices that induce reduced demand and increased supply or substitution. One implausible consequence of this theory is that energy consumption can be reduced arbitrarily with no implication or consequence for economic growth.

Future growth is simply assumed to be automatic, cost-free and independent of future energy costs. Thus the standard neoclassical economic theory is, in effect, 'dematerialized.' It needs to be 'rematerialized' in the sense of incorporating the laws of thermodynamics as real constraints on possible outcomes." Not coincidentally, Ayres is a physicist as well as an economist.

The notion that humanity will do fine or end up better off due to technological substitutes notably misfires when it comes to ecosystem services. Have scrubbers on coal-fired power plants made the air we breathe purer than it was before we ever started burning coal? Do captive breeding programs in 2005 produce levels of biodiversity that exceed the variety of wild animals that existed prior to industrial civilization? I can't think of many examples in which a technological substitution has equaled or improved an ecosystem service. And then there's the price tag. Take that colossal space mirror that could have cost up to $400 trillion to partially provide a service Mother Nature used to fully provide for free. Or, on earth, consider municipal water treatment plants. As I write this, city officials in a neighboring town of some 4,600 people are trying to figure out how to pay for a new treatment plant, which they estimate will cost about $9 million. That's around $2,000 for every man, woman, and child.

Carried to its extreme, the neoclassical production function implies that capital in the form of money can substitute for resources. To see how that notion has sometimes unfolded in the real world, let's take a journey to Nauru.

Nauru is the world's smallest island nation, a mere eight square miles of terra firma poking out of the South Pacific. For nearly all of its three-thousand-plus years as a home to humans—currently about eleven thousand of them—Nauru has existed in obscurity. In recent years, however, it has made headlines as the site of an infamous Australian-run refugee camp, whose conditions have been widely condemned as inhumane. Most of the condemnation targets the Australian government, but some has spilled onto the Nauruan government, as well, for agreeing to host the refugee camp in return for payments from Australia.

But Nauru's actions stem at least partly from desperation. Most Nauruans are poor and have few prospects for bettering their plight. At times the unemployment rate has exceeded 80 percent. Many suffer from ill health, notably diabetes and heart disease; according to the CIA's *World Factbook*, Nauru has one of the three highest obesity rates in the world. Yet just a few decades ago Nauruans were among the richest people on earth, with a per capita income that ranked among the top five nations.

It turns out that the island of Nauru largely consisted of high-quality phosphate, a valuable commodity and key ingredient in fertilizer. I use the past tense advisedly. Foreign mining interests took note, and during the 1900s they extracted many shiploads of phosphate. Nauruans made a lot of money, but by the end of the century most of the phosphate was gone and the strip-mining had devastated some 80 percent of the island and many of the surrounding coral reefs. "Pleasant Island," as an eighteenth-century English explorer had dubbed this once-fair land, had been largely reduced to a wasteland unfit for farming, building, tourism, or pretty much anything else.

Long aware that the phosphate would run out, Nauruans had followed the neoclassical game plan; as a substitute for their disappearing resources and ecosystem services, they invested their money. Long story short, most of their investments flopped. By the start of this century their wealth had largely vanished, the economy had cratered, and they had little natural capital to fall back on. As Ayres writes, "A substitution of natural for manufactured capital may be one-way:

once something is transformed into manufactured capital there is no way to return to the original situation."

Would substituting money for natural capital have worked out fine if the Nauruans had been wiser investors? Before answering, I should note that even if they had been brilliant investors, they might have fared poorly because they had the misfortune to run headfirst into an Asia-wide financial crisis. It goes to show that investing is inherently risky and not as solid as the land beneath your feet. Nauruans are hardly the first people to sell out their environment for cash and end up with neither a healthy environment nor a healthy bank account. That's a bit of financial realism that complements biophysical realism.

So let's rephrase the question: Would substituting money for natural capital have worked out fine if the Nauruans had been wiser investors *and* lucky and had raked in loads of money? It would have worked out better, no doubt, but far from fine. If they had loads of money, many Nauruans probably would migrate in order to escape their devastated island. But abandoning an ancestral home takes a toll as you lose your friends, your community, and your culture. If the Nauruans stayed and tried to compensate for the trashing of their island by using their money to buy stuff, they'd no doubt derive some enjoyment from those material things. But substituting stuff for ecosystem services and other public goods has its limits. As economist Peter Howard, the lead researcher for the Cost of Carbon Pollution project, asked, "If I lose the ability to go for a walk in the forest, can I just buy more iPods and get happier?"

Nauru's plight brings to mind our previous discussion of the precautionary principle. If only the Nauruans hadn't violated a key principle of precaution: don't count your chickens before they hatch. But like most of us, they were immersed in a growth-über-alles economy implicitly based on the eternal triumph of substitution, which fosters unwarranted optimism about compensating for lost resources. In hindsight, they might have done better if they had opened only a small portion of their island to mining as a pilot project, invested the proceeds, and waited to see what happened. How did the investments fare? How did the island's environment fare? How did the Nauruans' culture fare? In general, how did their experiment with substitution turn out? If it had turned out to be a net win, then they could have ramped up the phosphate mining. But when it began turning into a net loss, they could have stopped before they had ruined their land, the literal foundation of their nation.

Counting substitutes before they hatch could have global consequences in the case of climate change. Examples abound, but one of the most dangerous is the premature reliance on bioenergy with carbon capture and storage (BECCS). Leaning on this rickety concept injects unfounded optimism into climate calculations and policy at the highest levels, including Intergovernmental Panel on Climate Change reports and the Paris accord.

Here's how BECCS works: we grow plants on hundreds of millions if not billions of acres of land, those plants absorb vast amounts of carbon as they grow, we burn those plants to produce electricity, we capture the carbon emissions before they escape into the atmosphere, and we bury that carbon. Voilà! We have not only generated power without emitting greenhouse gases, but we have actually removed carbon from the atmosphere—so-called "negative emissions."

At least, that's how people hope BECCS will work. But so far the concept exists mainly on paper and has made little headway in the real world. Governments and businesses have spent many years and billions of dollars trying to develop BECCS, but a slew of engineering and financial problems have held it back. In addition, BECCS suffers from some existential flaws that may be insurmountable. For instance, to reach the necessary scale, we would likely have to grow plants on an area roughly the size of America's lower forty-eight states, give or take a billion acres. And we'd have to do this while also scrounging for more land on which to grow all the food we'd need to feed an expanding population.

Humanity is rolling the dice by counting significantly on the success of BECCS and other substitutes for achieving a stable climate. Many climate advocates worry that the seductive promise of negative emissions or any other supposed technological saviors might sap the urgency from efforts to curb emissions soon and drastically. Fossil fuel companies and their advocates already are citing BECCS as an excuse to keep expanding their operations for another decade or two or three. Also hazardous is the fact that the rest of us are unwittingly supporting delays in curbing emissions because an unearned confidence in substitutes is baked into calculations on which we base many of our climate decisions.

WATER IS A GIRL'S BEST FRIEND

Why do diamonds cost more than water? This is the classic diamonds-water paradox, pondered by economists at least as far back as Adam Smith in the eighteenth century. We need water to live, yet we pay far more for diamonds, which we don't need. Why? For one thing, water is plentiful and diamonds are scarce. Of course, if you're dying of thirst in a desert, you'd trade diamonds for water because in that scenario both commodities are scarce, but the water is a necessity. So it is scarcity combined with demand that determines value, and the biophysical reality that you need water to survive can sometimes create a demand that trumps all the diamonds in the world. In such a situation, neither a diamond nor the monetary worth it represents can substitute for water.

Likewise, neither money nor market goods can substitute adequately for many nonmarket goods, notably ecosystem services. This was an abstract point in preindustrial times because humanity had little or no impact on services such as pollination, nutrient cycling, and climate stability. They were abundant and free and had no exchange value, particularly compared with human-made goods that were scarce and in demand, like a spear or a pelt.

But as the centuries passed, the scarcity balance started turning, slowly at first but with increasing speed as the Industrial Revolution accelerated. These days the scales have shifted so far that many ecosystem services are rarer than many human-made goods and services, and getting rarer all the time. The demand for these increasingly scarce services is rising and therefore their value and implicit price should likewise rise.

Adjusting climate IAMs to reflect the growing scarcity of ecosystem services and other nonmarket goods would dramatically raise the SCC. Two Swedish professors, Thomas Sterner and Martin Persson, explored this idea in a study in 2008. They point out that "implicit in all integrated assessment models (IAMs) used in the analysis of climate change policy . . . lies the assumption of perfect substitutability. Perfect substitutability implies that detriments of climate change impacts can be balanced with increased consumption of material goods on a one-to-one basis: one dollar's worth of climate damages, regardless of the kind, can be compensated by a dollar's worth of material consumption, so that despite climate impacts we will be richer and enjoy a higher level of welfare in the future." But we can't replace coral reefs with big-screen TVs. As ecosystem services become less common due to climate change, they will become more and more valuable and their loss will grow more and more expensive. By mistakenly assuming that human-made goods can substitute for these declining necessities, IAMs greatly underestimate the cost of the damages that global warming will cause.

The increasing scarcity of ecosystem services and other nonmarket goods has implications that reach far beyond the SCC and climate change. Metaphorically speaking, we should redefine "growth" to mean growing water, not diamonds. The implications reach further yet when we expand these principles beyond the environment to the entire economy and its purpose. What is scarce and valuable, with "valuable" defined broadly as that which makes us happy? Is our economy delivering what is most valuable?

Consider the following excerpt from a former U.S. president's speech that touches on this broader idea. See if you can guess which president delivered these words:

The purpose of protecting the life of our Nation and preserving the liberty of our citizens is to pursue the happiness of our people. Our success in that pursuit is the test of our success as a Nation.

For a century we labored to settle and to subdue a continent. For half a century we called upon unbounded invention and untiring industry to create an order of plenty for all of our people.

The challenge of the next half century is whether we have the wisdom to use that wealth to enrich and elevate our national life, and to advance the quality of our American civilization.

Your imagination, your initiative, and your indignation will determine whether we build a society where progress is the servant of our needs, or a society where old values and new visions are buried under unbridled growth. For in your time we have the opportunity to move not only toward the rich society and the powerful society, but upward to the Great Society.

The Great Society rests on abundance and liberty for all. It demands an end to poverty and racial injustice, to which we are totally committed in our time. But that is just the beginning.

The Great Society is a place where every child can find knowledge to enrich his mind and to enlarge his talents. It is a place where leisure is a welcome chance to build and reflect, not a feared cause of boredom and restlessness. It is a place where the city of man serves not only the needs of the body and the demands of commerce but the desire for beauty and the hunger for community.

It is a place where man can renew contact with nature. It is a place which honors creation for its own sake and for what it adds to the understanding of the race. It is a place where men are more concerned with the quality of their goals than the quantity of their goods . . .

Will you join in the battle to build the Great Society, to prove that our material progress is only the foundation on which we will build a richer life of mind and spirit?

There are those timid souls who say this battle cannot be won; that we are condemned to a soulless wealth. I do not agree. We have the power to shape the civilization that we want.

So, who was the speaker? Most of you probably noticed some anachronistic language and ideas in the speech, such as the sexist use of "man" and "men" or the implicit colonialism in the phrase "settle and subdue a continent." These suggest a president from an earlier era (though not necessarily). Many older readers and those with a solid knowledge of American history likely picked up on the dead-giveaway clue: the repeated use of "Great Society," which was the name of the signature series of domestic programs launched by the answer to our quiz: Lyndon Baines Johnson.

I was surprised that LBJ gave this speech, considering his reputation as a crude, bare-knuckled, and sometimes deceitful politician. But more surprising to me is the fact that he gave this speech in 1964, before the environmental movement blossomed and the concept of limits to growth became more common. Most surprising of all is the fact that any American president ever gave such a speech, considering that the glory of growth has long been an article of faith to most Americans.

Johnson proclaimed that the challenge of the next half century was to focus more on the quality of our lives than the quantity of our stuff. More than half a century has passed since he gave his speech, and we have not met that challenge. Yet the imperative has only grown clearer and greater.

11

JUDGMENT DAY

"We have the power to shape the civilization that we want."

If this LBJ line that I quoted near the end of the previous chapter sounds like a sentiment from the kind of aspirational speech native to graduation ceremonies, that's because it is. Johnson was exhorting the class of 1964 at the University of Michigan. Cynics may arch a world-weary eyebrow at such effusions, but beneath LBJ's rhetoric lies an intimidating but exhilarating truth that we must take to heart in order to advance a sustainable, just, and prosperous economy: we, the people, do indeed possess civilization-shaping power.

God did not hand down the market from on high. Nor is it an immutable natural phenomenon, like the rotation of the earth. People built the current economic system, and people can fix it. We should not cede our power and shirk our responsibility by blindly following the mechanistic decisions of a supposedly omnipotent market. Ours is more of a God-helps-those-who-help-themselves situation. We need to fire up our free will and use it to create the political will to do what needs doing.

Sounds good. One thing, though. What is "political will," exactly?

Ah, yes, political will. This term shows up at the end of countless articles, presentations, and books that suggest ways to deal with climate change and any number of other big, thorny issues. If you play a drinking game that requires taking a swig every time you encounter that term, you'd better not plan on driving yourself home. Typically these writings offer all sorts of compelling solutions, including many specific actions that are eminently doable. But then you reach the conclusion and the author pops your balloon with the almost obligatory comment about how we could do all these wonderful things if only we could summon enough political will.

Consider the comments of UN Secretary-General António Guterres, speaking in late 2019 in the lead-up to a big international climate conference. He noted that we have the scientific and technical means to address climate change and yet our efforts have been "utterly inadequate." He was quite clear about the fundamental problem. "What is lacking is political will. Political will to put a price on carbon. Political will to stop subsidies on fossil fuels. Political will to stop building coal power plants from 2020 onwards. Political will to shift taxation from income to carbon."

A few years ago Lori Post, Amber Raile, and Eric Raile wrote a scholarly paper titled "Defining Political Will" that does a credible job of bringing this elusive term into clearer focus. Their bare-bones version defines political will as this: "The extent of committed support among key decision-makers for a particular policy solution to a particular problem." The paper fleshes out this spare definition with detailed dissections of these factors, such as the importance of the intensity level of the decision-makers' commitment to a policy solution and the difficulty of reaching a widely agreed-upon interpretation of the problem or a shared vision of the solution. But you don't have to machete your way through every thicket in the paper to see how tough it can be to gather enough political will around any particular solution to any particular issue, especially in a polarized political atmosphere like America's.

Though it would still be tough going, you can find a path to political will in "Using Public Will to Secure Political Will," a paper by Post, Amber Raile, and Charles Salmon that dovetails with "Defining Political Will." This complementary paper makes a crucial point: building committed support among decision-makers often requires committed pressure from citizens. The power of public will presents us with great opportunity.

This book has now reached its own "political will" juncture: I could provide a chapter or two full of useful actions, mutter about the need to muster political will, and then say adios. But given that we're beset by environmental and social crises, I think we should try to go a little further. So I will provide a few examples of actions, but I'll forgo the customary long lists. Instead, in these last two chapters we'll explore some of the overarching attitudes and misperceptions that make it difficult to assemble political will, and we'll consider some ways we might overcome those obstacles.

Let's begin by diving deeper into the idea of free will in the realm of economics. The disputes over the social cost of carbon provide a ready example of the wrestling match between those people who want to leave decisions largely in the invisible hands of the market and those who want to use our own hands to more purposefully mold our economic future. Setting aside the debates over the calculation and use of the SCC, the decision to develop an SCC at all signals recognition of the market's failure regarding global warming and of the need for us humans to step up and make some changes. This is one of the reasons that fans of the invisible hand (and the predators who hide behind the invisible hand) have been working to gut the SCC.

Yet the SCC is not dead, though as of 2020 the Trump administration has executive-ordered the federal SCC into a coma. Other nations and some U.S. states are using an SCC or something similar. For example, Minnesota has been using an SCC for years when making utility decisions, albeit a very low figure that for many years ranged from 44 cents to $4.53 per ton of carbon dioxide. In 2014, the public utility commission greenlighted the state's largest solar project instead of a competing natural gas plant even though by orthodox calculations the gas project would have been cheaper. But via Minnesota's SCC the commission accounted for climate change externalities and that helped show that solar was actually the less costly choice. And that decision was made with the commission using that paltry SCC. In 2017, the commission decided by a 3–2 vote to dramatically raise the SCC range to $9.05 to $43.06. (The two dissenting commissioners wanted an even higher SCC.) Those new numbers will greatly bolster the economic case for climate-friendly decisions.

The SCC continues to live in the halls of justice, too. Some courts have recognized the pre-Trump federal SCC as valid. In 2016, in a pivotal case regarding energy efficiency standards, a panel of three judges (all appointed by Republican presidents) of the U.S. Court of Appeals for the Seventh Circuit declared that the Department of Energy was reasonable in applying the SCC. This was the first time a higher court had made a ruling about the legality of the federal social cost of carbon. That set a positive precedent, but the SCC's legal future is still hazy given the antagonistic actions of the Trump administration.

The best hope for the federal SCC lies in the 2020 elections. However, even if supportive politicians take power in D.C. and champion the social cost of carbon, how they will put it to work is yet to be determined. As we've seen in this book, the right-wing detractors are not alone in expressing qualms about the SCC. Many scientists, climate advocates, and sustainability economists also express qualms, albeit different ones, fearing that the SCC's flaws might be overlooked and that policy makers might misuse it.

These climate-friendly critics suggest a variety of ways to properly use an SCC, many of which share an underlying theme: the SCC and climate economics in general should play a vital but secondary role. Instead of looking to the SCC and economics to tell society how much to spend combating climate change, these critics urge us to let biophysical reality be our primary guide. They counsel setting a science-based target, building in a serious buffer as insurance against fat tails, and having governments make sure we hit that target.

For example, based on the best science, we might decide that we should hold the temperature rise under 1.5°C above preindustrial levels. Following that decision, economics, perhaps including the SCC and carbon pricing, could enter the scene to help determine the most efficient and equitable way to achieve sufficient greenhouse gas reductions to hit that target. "Economics would find itself in a humbler role, no longer charged with determining the optimal policy," write the authors of a study that examines the science-first idea in the context of risk. Once science has indicated safe emissions levels, "there remain the extremely complex and intellectually challenging tasks—for which the tools of economics are both appropriate and powerful—of determining the least-cost global strategy for achieving those targets."

However, though biophysical reality has a huge role to play, by itself it can't determine our climate change policies any more than economics can. Only people can choose which biophysical reality fits our wants and needs and then strive to achieve that reality. Take the above example of keeping the planet's temperature rise under 1.5°C. Why not 2°C or 3°C? Biophysical reality has no preference. Many advocates for rigorous climate action support 1.5°C because the best available science indicates that we humans probably would be able to live and even prosper in the resulting world. So it is biophysical reality filtered through the lens of humanity's desires that has led to a widespread agreement that 1.5°C would be acceptable. It is a choice deeply informed by science but made according to our judgment as to how to best serve our values.

There, I've said the j-word: "judgment." Hard-right economists and their followers would like to largely banish judgment from the economy, aside from consumers and producers making the quotidian judgments that fuel supply and demand. Free enterprise purists view most conscious efforts to shape the market, aside from minimal rules of the road, as central planning that will foul the gears of the economy. In *The Constitution of Liberty*, Friedrich Hayek, a deity of the conservative economics pantheon, urges people to submit to "an impersonal mechanism [the market], not dependent on individual human judgments."

Hayek's impersonal mechanism serves an equally insensate goal much sought after in orthodox economics: efficiency. Sure, efficiency is desirable, but it's not a goal. It is a means to an end. Venerating efficiency as the overarching purpose of the economy makes no sense to sustainability economists. They would ask, "Efficient at doing what?" They know that a hammer can be used either to efficiently build a table or to efficiently smash a table into splinters. Deciding what you want to efficiently do with that hammer requires judgment.

In a paper entitled "Envisioning Shared Goals for Humanity," economists Josh Farley and Robert Costanza write, "First and foremost, the economist must decide what ends are to be pursued." Or, as dugout sage Yogi Berra supposedly put it: "If you don't know where you're going, you'll end up someplace else." Sustainability economists believe we should set societal goals and then fashion an economy that will get us there.

The role of human judgment in the economy is controversial, but the last sentence in the above paragraph contains two words that are downright incendiary in the context of economics: "we" and "us." These words refer to people making a collective judgment, and the idea of collective judgment brings us to . . .

GUB'MINT

"You favor a woman's right to abortion but not a woman or man's right to a light bulb," accused Rand Paul, the Republican U.S. senator from Kentucky.

"Huh?" you might well ask. I reckon a little background on Light Bulb War I is in order.

This conflict began in response to a minor element of an energy bill passed by Congress and signed into law in 2007. That element enacted energy-saving standards for light bulbs that would ratchet up and eventually be strict enough to effectively ban old-fashioned incandescent bulbs. "I find it insulting," said Paul. He characterized the federal bureaucrats implementing these rules as "busybodies" trying to tell Americans how to live their lives. His outrage was directed at a Department of Energy official who was testifying that more efficient light bulbs would both save consumers money and benefit the environment.

Paul was not standing alone on the ramparts, as the mercurial libertarian sometimes does. A fair number of congressional Republicans enlisted to fight the invasion of efficient bulbs. Some even pushed legislation aimed at preserving Americans' God-given right to use inferior light bulbs. For example, GOP representative Joe Barton of Texas sponsored the Better Use of Light Bulbs Act in an effort to roll back the energy-saving standards. He worried that America would lose the source of light that, as he put it, "has been turning back the night ever since Thomas Edison ended the era of a world lit only by fire in 1879." With the same intent, Republican representative Michele Bachmann wrote the Light Bulb Freedom of Choice Act. "Instead of a leaner, smarter government, we bought a bureaucracy that now tells us which light bulbs to buy."

Incidentally, if Bachmann's line sounds familiar, it may be because GOP politicians and conservative pundits tossed around similar notions in 2019 when Light Bulb War II broke out, ignited by the Trump administration's rollback of light bulb efficiency standards. While making various disparaging—and inaccurate—remarks about efficient light bulbs at a GOP retreat, Trump also said this: "And I looked at it, the bulb that we're being forced to use, number one to me, most importantly, the light's no good. I always look orange." Perhaps the most bizarre assault of Light Bulb War II came courtesy of Fox News' Laura Ingraham. Her combo attack incorporated incandescent light bulbs with two other supposed liberal bugaboos: beef and plastic straws. During an on-air segment she tried to suck a steak stuffed with incandescent light bulbs through a plastic straw. I'm not sure what to say about that.

Before we turn off the lights on this burning issue, let's return to Light Bulb War I and Representative Ted Poe. In a fiery speech on the House floor, the Republican from Texas lambasted the new standards while holding aloft an example of what was in that pre-LED time the hated enemy: a compact fluorescent light (CFL) bulb. "The bill does one thing, Madam Speaker," Poe thundered. "It controls the type of light bulbs that all Americans must use throughout our fruited plains." At one point he put down the CFL and brandished a copy of the Constitution. "I don't see anywhere in the U.S. Constitution that it gives the government the power to control the type of light bulbs used in Dime Box, Texas, or any other place in the United States." After a long rant about CFL disposal, he asked sarcastically: "Will the EPA light bulb police haul us off to jail because of improper disposal procedures?" He wrapped up with a theatrical flourish: "Oh, I yearn for the day when America took care of Americans by developing our own abundant natural resources, like coal and natural gas and crude oil to provide affordable energy to Americans. But those days have gone the way of Edison's incandescent light bulb. We might as well turn out the lights. The party's over."

"The party's over." A telling turn of phrase. The unregulated society Poe and company yearn for does indeed bear a close resemblance to some anarchic bash. Think unsupervised adolescents and a keg. No one to make sure they don't drive drunk or commit sexual assault. Most of the teens would handle their freedom well and avoid any serious transgressions, but some not so much. This latter group writ large is one of the main reasons we created laws in the first place. And in today's complex world the need for laws has multiplied, most definitely including those that guide the economy.

Which is why we need to talk about gub'mint. I use the slangy spelling to convey the antipathy so many Americans have for government. That this antipathy has been growing in recent years is no accident. For several decades organized interests with deep pockets have been preaching deregulation and belittling bureaucrats in a largely successful campaign to turn "government" into "gub'mint." Ronald Reagan summarized this sentiment with a famous line in his first inaugural address, in 1981: "Government is not the solution to our problem. Government is the problem."

So-called public choice proponents provide an example of the anti-gub'mint effort. They assert that the dominance of government by predators is inevitable in a democracy. This expectation of dominance by self-serving interests has led public choice scholars to proclaim that market failures are a lesser evil than "government failures," the term they apply to what they picture as widespread, inevitable, and unmanageable government corruption. For example, they declare that regulatory agencies such as the EPA are and will forever be captured by special interests, such as environmental groups or the industries the EPA is supposed to oversee, so most public choice theorists champion less regulation—a lot less.

This posture is a bit rich, however, given that public choicers typically hail from the end of the political spectrum inhabited by predatory industry interests, who have engineered a number of so-called government failures. Take that concern about the capture of regulatory agencies. Recent GOP presidents, Trump emphatically included, have energetically promoted such capture by packing regulatory agencies with appointees with close ties to the industries they are supposed to supervise. As intended, this has led to "government failures" that profit the predators while costing the rest of us. A few public choice purists may criticize such self-serving abuses of government power even though Republicans are the greatest sinners, but criticism of conservative politicians and the companies they coddle is uncommon in the public choice movement.

It turns out that the influence on government that most public choicers truly fear is that of what they might term the masses. Beneath the garden-variety hypocrisy of much public choice work lies a corrosive, profoundly undemocratic ethos, as historian Nancy MacLean chronicles in her book about public choice, *Democracy in Chains*. The movement's embrace of the term "public" is misleading because their theory exudes contempt for the public. A public choice most-wanted poster would not depict a rapacious coal baron or a scheming Wall

Street financier. It would show a public school teacher lobbying for more spending on education or an elderly man cashing his Social Security check. Public choice writings and speeches bristle with complaints about the unproductive, mooching majority using its voting power to bully politicians into taxing the so-called makers (wealthy business owners and executives) to give to the so-called takers (most of the nonwealthy public) through government programs.

MacLean focuses much of her book on the founding father of public choice theory, economist James Buchanan (not to be confused with America's fifteenth president). According to MacLean, Buchanan referred to people benefiting from tax-transfer programs as "parasites." And he wasn't just concerned about traditional welfare programs. He saw Social Security, Medicare, public education funding, environmental protection, and almost any other form of social spending as undeserved government handouts that intrude on the liberty of the propertied and that should be eliminated. Perhaps not incidentally, such actions would result in massive tax cuts for the wealthy, big profits for the private sector, and rampant deregulation.

Public choice theory acknowledges only one kind of making and that is the making of money. If you don't earn enough, you have only yourself to blame. Buchanan and other hardcore public choicers don't want to hear any excuses about bad luck, poor health, or being born into circumstances that make amassing wealth almost impossible. And they especially don't want to hear complaints about the economy being rigged to favor the powerful and wealthy.

Buchanan's narrow definition of productivity excludes all other forms of making, such as making the world a better place. From our discussions of nonmarket goods we know that money is the measure of what the markets value but not necessarily what we humans value. Picture a sixty-five-year-old woman on the verge of retirement. Rather than having chosen to single-mindedly rake in as much money as possible by any means necessary, maybe she opted for a career as a family farmer, a doctor in a free clinic, or an army chaplain. She might have been great at one of these important but modestly paid jobs, but perhaps she was not skilled at managing money. Or maybe she simply didn't want to put her time and energy into managing her money. Or maybe she spent much of her retirement money supporting aging parents. Whatever the reason, when she retires, she's going to need a decent-sized Social Security check. Buchanan would label her a taker despite her decades of good work. On the other hand, Buchanan would label someone a maker if that person retired with a multimillion-dollar nest egg, even if that nest egg was the result of a lifetime of squeezing minimum-wage employees and skirting environmental laws. Public choice theory promotes a pinched, self-centered concept of responsibility.

I think most people reject Buchanan's sad view of what constitutes a maker. My own experience leads me to believe that most people are generous at heart. (I like to think that includes most public choice advocates, though their generosity seems focused on private charity as opposed to an open-armed embrace of the public good.) I came across an odd bit of evidence that supports my belief. If you're wondering just how important it is to people to help others, take a look at obituaries. Lux Narayan did. The CEO of a social media intelligence company, Narayan analyzed about two thousand *New York Times* obits to search for patterns in the content. He discovered that by far the most common word used in loving descriptions of the careers and lives of the deceased was "help," "helped," or some other form of that revealing word.

Let's shelve the hypocritical and antidemocratic elements of public choice theory and simply consider its assumption that ambitious government inescapably leads to rampant predation and does more harm than good. I think most of us would agree that too many predators stalk the halls of government, but public choicers willfully overlook the huge amount of good that our government manages to do despite being hobbled by public choicers and other gub'mint haters.

Consider the actions of Frances Kelsey. A family doctor and professor of pharmacology, Kelsey, in 1960, took a job with the Food and Drug Administration (FDA) evaluating new drugs. One of the first applications she reviewed came from the William S. Merrell Company, a major pharmaceutical outfit. Merrell was seeking a license to sell a drug called Kevadon in the United States. It was already used widely in Europe to help pregnant women with morning sickness, so it seemed certain to easily pass muster and quickly become available to Americans. But some of the information Merrell provided struck Kelsey as questionable, so she asked the company for more data.

Merrell was not pleased at the delay. They figured Kevadon to be a big earner, and they'd already stockpiled tons of it in warehouses and sent samples to a thousand doctors. A yearlong battle between Kelsey and the pharmaceutical giant ensued. As she asked for more and more information, the company resisted, branded her a meddling bureaucrat, and went over her head at the FDA to try to force her to support the license application. But Kelsey and her team discovered more and more problems with Kevadon and refused to give in to Merrell's bullying—and her FDA superiors backed her up.

Then, in 1961, the heartbreaking reports from Europe and other countries began rolling in. Kevadon, which we've come to know by its now-notorious generic name, thalidomide, was causing horrible birth defects. Tens of thousands of babies were born with a range of problems, including deformities of their eyes, an absence of external ears, and flipper-like appendages instead of arms and legs. This meddling bureaucrat and her colleagues spared untold numbers of American babies and families from the tragedy of thalidomide. President John F. Kennedy made Kelsey a key figure in his successful effort to pass a landmark law tightening drug regulation. She went on to a stellar forty-five-year FDA career, part of it as director of the FDA's Office of Scientific Investigations, where she continued to champion drug safety.

Much of the good that government does takes the form of fixing and preventing the abundant market mishaps that plague us, such as side effects of thalidomide. I seriously doubt that government failures cause more harm than existing market failures, let alone more than the mountain of additional market failures that would pile up if government regulators backed off as far as public choice economists favor. And what if we had reasonably strong, honest government that excluded most predators? I think public choice proponents grossly underestimate the vast number of externalities that could be eliminated or at least greatly decreased by good governance. Climate change comes to mind.

We need to drop our knee-jerk hostility toward the idea of government in general and instead direct our hostility toward *bad* government. Sure, we've experienced way too much bad governing. Yes, in some cases less government is the answer. But in some cases more government is the answer. And in some cases different forms of government are the answer. The basic point is that we need good government, whether the situation calls for less, more, or different. And good government will only happen if enough people do the job of being good citizens, which requires individuals to choose to work together and take collective action.

"[Climate change] is a case where you need collective action and inevitably that's a political process," said Dean Baker, the cofounder and senior economist at the progressive Center for Economic and Policy Research. "Most economists want to leave it to the market." But, said Baker, it takes government to accomplish many of the actions required to deal with climate change and to achieve sustainability in general, such as zoning for compact communities and boosting public transit. "Government actually does these things already, but we just don't think about them," said Baker. Baker added that often collective action is simply cheaper and more efficient. For example, having city government establish a fire department works far better than having individuals hire their own firefighters.

Think of collective action via government on behalf of one's country as an act of patriotism. Look back to the formation of the United States of America, in the fall of 1787, when the people of our newborn nation were pondering whether to adopt the Constitution. (Well, when propertied white males were so pondering—democracy had a long way to go, as it still does.) In the first of the *Federalist Papers*, in which Alexander Hamilton, James Madison, and John Jay argue in favor of ratifying the Constitution, Hamilton emphasizes the monumental decision regarding government that America's freshly minted citizens were about to make. In the first paragraph, Hamilton writes: "It has been frequently remarked that it seems to have been reserved to the people of this country, by their conduct and example, to decide the important question, whether societies of men are really capable or not of establishing good government from reflection and choice, or whether they are forever destined to depend for their political constitutions on accident and force."

If the idea of acting collectively still rubs you the wrong way, perhaps you wouldn't chafe so much if you think of it in analogies. Being an engaged citizen doing your part to help government function well is like being an engaged member of a basketball team doing your part to help your team play well. Or like being in the military and doing your part to help carry out the mission. Or like belonging to a tight-knit farming community and doing your part to help with a barn raising. We individuals often act collectively. Not only is it the most effective way to do many things, but we relish the camaraderie, the sense of pulling together, and the close bonds we form with the people with whom we work to get things done. Choosing to act collectively when it's the best way to do the job does not constrict our freedom and self-reliance; it expands them.

Bear in mind that "government" is a supple term that extends well beyond the Washington, D.C., Beltway. It certainly includes state and local governments, where we non-power-elite folks can more readily work up close and personal with elected officials and public employees to get things done. Consider an example from Maryland.

Some years ago Maryland state economists figured that its coastal wetlands provided a variety of values, including non-market services. That's admirable but not revolutionary, as you know from our earlier discussion about assigning a rough monetary value to ecosystem services. What's most notable is the next step Maryland took. In 2010, Maryland became the first state to adopt a genuine progress indicator (GPI).

The GPI is one of many methods of measuring welfare that are vying to replace or at least supplement the GDP, which merely measures economic activity. The GDP goes up when a couple has a baby and pays the obstetrician. The GDP also goes up when someone is murdered and the bereaved family pays the funeral home. GDP makes no judgment as to whether the occurrence that generated the economic activity adds or subtracts from the well-being of society. The more killing sprees, the better the GDP looks. Many critics have denounced the use of the GDP as a measure of how well we're doing—even its creator warned people not to use it as a proxy for social welfare—yet many people keep using it that way.

Abiding by the GDP has profound implications, too, because it both reflects and affects our deep economic thinking. As the old expression goes, you get what you measure.

But the people in Maryland wanted a more meaningful metric, so groups of stakeholders and experts gathered and hammered out a list of twenty-six social, economic, and environmental qualities that they considered vital indicators of the health of society. These markers ranged widely, from hours spent watching television to the amount of coal burned. A few years later Maryland expanded the list to fifty items, including seven ecosystem services, some of which coastal wetlands provide abundantly. So when the state considers which lands to buy or which restoration projects to undertake, the planners include ecosystem services and other nonmarket values in the decision-making.

Elliott Campbell, who oversees Maryland's GPI, told me of an example that demonstrates how ecosystem service values can affect the state's economic decisions. The Department of Natural Resources wanted to buy and protect a couple of hundred acres of wetlands on Maryland's eastern shore, but the Board of Public Works, which must approve land acquisitions, balked at the price tag. However, said Campbell, "We were able to show that the wetlands had a tremendous amount of ecological value for things like groundwater recharge and wildlife habitat and carbon sequestration." Sold! The board gave a thumbs-up, and the state bought the property.

Maryland's GPI process shows what collective judgment can look like. People got together and made choices. They discussed their values, figured out goals, and formed plans. This provides an example of how people making value-based judgments can free us from the narrow confines of the market and enable nonmarket goods to compete with private goods.

Bringing stakeholders together, early and often, is particularly important in order to achieve environmental justice. For collective judgment to work it must collect the judgments of everyone involved, but too often people of color and low-income residents get left out even though they disproportionately suffer from environmental harms.

The need to include all stakeholders definitely applies to the climate plans that are cropping up in cities all over the United States. It's heartening to see so many ambitious plans being adopted, and they're getting better at including equity issues, but most still fall short. According to the American Council for an Energy-Efficient Economy, only twenty-four of seventy-five cities surveyed address equity in climate and energy planning, and only a handful earn high marks. A 2019 *Grist* article by Zoe Saylor, titled "Providence Shows Other Cities How Environmental Justice Is Done," makes a convincing case that Providence, Rhode Island, arguably serves as the best role model. Not only does the city's climate plan deal with specific problems, such as the industrial waste and polluted air that poison marginalized neighborhoods, but to ensure serious input from these communities, the city is instituting a collaborative governance model. This requires the Office of Sustainability to work together with people of color and low-income residents on all its projects. Such collaboration reveals issues that otherwise might be missed, such as the concerns that cleaning up Providence's industrial sites and building new transit opportunities would initiate "green" gentrification. But thanks to an inclusive process, the city heard this concern and took steps to provide affordable housing to help low-income residents stay in their neighborhoods.

The term "government" reaches beyond states and cities, too. It includes any form of collective decision-making and management that features some element of enforcement.

Commons provide a classic example. Many people know the term "commons" from Garrett Hardin's famous essay "The Tragedy of the Commons," published in 1968. Hardin describes how shared resources that are available to all comers can be overused by individuals acting in their "rational" economic self-interest, resulting in a damaged resource of little use to anyone. The overfishing of the bluefin tuna, recounted in an earlier chapter, provides a textbook example.

But commons do not have to end in tragedy. Hardin recognized this, but people often misconstrue his work to mean that commons inevitably lead to overexploitation and destruction. Government agencies can preserve commons, as various American federal and state departments do, however imperfectly, through control of resources such as forests and groundwater. But such management can also occur via homegrown governing bodies. Elinor Ostrom spent decades demonstrating that in some cases commons could be sustainably managed by networks of local resource users, research for which she won the Nobel Prize in Economics in 2009. She cited many examples of traditional systems that have been working for a long time, such as schemes for grazing common pastures in the Swiss Alps that date back to the sixteenth century. Whether management by local users or by a central government works better depends on the circumstances, but some kind of governance is needed.

FREEDOM

Milton Friedman and his wife, Rose, also an economist, wrote a bestselling book called *Free to Choose*, published in 1980. The first word in the title conveys an overriding theme in the conservative assault on government regulation: unbound markets are essential to freedom, and government interference in markets is anathema to freedom—and not just to economic freedom but to political freedom. In the book the Friedmans write that "an even bigger government would destroy both the prosperity that we owe to the free market and the human freedom proclaimed so eloquently in the Declaration of Independence."

Conservatives have branded government intervention in the market as a blow to freedom by equating most economic planning with Soviet-style micromanagement, in which a centralized bureaucracy tells a business how many blue and how many green widgets to produce instead of letting businesses take their cues from consumers freely exercising their preferences for blue or green. But this specter of Soviet-style intervention is a straw man that is not being promoted by the advocates of good government. This definitely includes sustainability economists. A key principle of sustainability economics is to strive for as little regulation as possible.

This key principle raises a key question: How little regulation is possible if we want to have a sustainable, just, and prosperous economy? Totally unrestrained behavior is simply anarchy, in which a big guy with a big club can exercise his freedom by bashing you over the head and taking your food. Not much freedom for you. Even the most rabid market fundamentalists reject such a chaotic vision and recognize the need for basic laws and standards of conduct. However, in the economic arena, they generally emphasize laws and standards that protect private property, which leaves many non-market goods unprotected. Such a weakly regulated system often results in one person's economic freedom infringing on the freedom of another person via externalities. Lyle Perman, a fourth-generation rancher in South Dakota, captured the essence of the idea in a comment he made to *The American*

Prospect in 2014: "I don't like the government telling me what to do," he said. "But if your actions impact somebody else, then it becomes somebody else's business, too. And that's where I draw the line."

A 2016 paper from the Center for Progressive Reform offers a passage that neatly summarizes the gub'mint issue: "For decades, we have been stuck in a debate based on the false premise that markets enhance liberty while government restricts it. That is wrong. Properly organized, government becomes an agent of liberty by promoting opportunity, health, and security for all. It is hard to fully enjoy your liberty when a preventable workplace accident puts you in a hospital bed or your children cannot drink water from the tap. Debates about regulation must recognize that the appropriate question is what the mix of markets and government should be, not that markets are always good and government is always bad." This blend of private and public is called a mixed economy, and it's as familiar as sunshine in summer. All advanced economies are mixed, including that of the United States. It's just that America's blend leans to the market end of the spectrum.

In a sustainable economy the mix would be better balanced, with enough government involvement to take care of nonmarket goods. A lot of commerce would look much the same as it does today: a new restaurant would open on the riverfront; a car company would promote its latest model; a developer would build a residential subdivision. Government would have a hand in these enterprises, just as it does now. Health inspectors would certify the sanitation procedures of the restaurant, new cars would have to meet safety standards, and the houses in the subdivision would need to measure up to building codes. Also like now, some level of government would run or oversee certain services deemed essential, such as many utilities and most schools.

Many of the changes brought about by sustainability economics would be a matter of degree. That subdivision would already be subject to some environmental standards, such as not dumping toxic waste into a creek. In a sustainable economy the government would strengthen such rules and enforce them more vigorously. Government might also enact additional regulations for that subdivision, like mandating that the houses feature passive solar design and that neighborhoods have bioswales to naturally filter pollutants from runoff. When possible, government would avoid microregulating by going upstream and making changes at a higher systemic level. Perhaps the state legislature would pass a law requiring that all new houses meet zero-net-energy standards by 2025 and leave it up to the subdivision's developers and the market to figure out how to hit that target. Government already uses this approach fairly often, but a pervasive sustainability economics perspective would make such requirements more common. As I said, it is a matter of degree, though in some cases it would be a matter of many degrees.

Qualitative changes would also be needed to achieve a sustainable economy. Some would be specific, such as establishing banks that put people and place ahead of maximizing profits. Others would sprawl, such as the widespread institutionalization of our values in economic decisions, like the people and government of Maryland did in a small way with their genuine progress indicator.

Sustainability economics recognizes the complexity of freedom, but many market champions push a simple adolescent view that essentially defines freedom as "everybody get out of my way so I can do what I want." Unlike Perman, they fail to see that blithely asserting their freedom can interfere with other people's freedom. Let's say that somebody's great-granddaddy bequeathed him a tract of creek-side forest, and now he wants to clear it and build a subdivision, but environmental restrictions stand in his path. Frustrated, he heads to his favorite bar, where he nurses a beer and grouses about pointy-headed bureaucrats. But what he may not understand is that his freedom to develop his property conflicts with his neighbors' freedom to live without the noise and traffic the development would bring; it conflicts with anglers' freedom to catch the fish that would be decimated by logging the trees along the creek bank; it conflicts with the freedom of down-stream residents to live without the floods that replacing for-est with housing would magnify; it conflicts with everyone's freedom to live without the burden of climate change, which would be aggravated by the greenhouse gases released by the logging and the increased traffic; and so on.

Sustainability economics seeks to balance competing free-doms, which often requires using our collective judgment through government. As Barack Obama said in 2016 when speaking about the Flint, Michigan, drinking-water crisis: "This myth that government is always the enemy—that for-gets that our government is us, that it's an extension of us, ourselves. That attitude is as corrosive to democracy as the stuff that resulted in lead in your water."

Often conflicts involving competing freedoms arise because the human footprint has grown so vast. If there were no subdivisions in a watershed, and only relatively few in the nation and the world, the addition of one on your great-granddaddy's land probably wouldn't have a noticeable effect on other people's freedoms. Likewise, you could blithely send carbon dioxide up the chimney, log old-growth trees, and catch as many fish as you want anywhere you want any-time you want—none of these acts would infringe on anyone else and therefore require regulation if our society had never expanded on such a massive scale. Ironically, the people who complain the loudest about regulations often are the same people who promote the kind of heedless growth that has made so much regulation necessary. Equally ironic, if we supported good, active government and established a sustainable economy, we would be able to get rid of a lot of regulations because they would no longer be needed.

And who doesn't wish for less regulation? I wish wastewater could be channeled directly into the nearest creek so I wouldn't have to help pay for my town's sewer system and water treatment plant. However, I don't mind this minor expense compared with the major disaster of the polluted river we'd have if I and everyone else in town dumped our untreated wastewater into the closest stream.

Contrary to the assurances of *Free to Choose*, overreliance on the market limits our roles as citizens and confines us to our roles as consumers. Because the market emphasizes material goods and services for individuals, it doesn't give us the freedom to easily choose nonmarket goods and services, including many that we might consider far more important than the stuff the market makes it all too easy to spend our money on. Via government we can include these nonmarket items among our choices and enjoy more freedom, not less.

Counterintuitively, underregulation also restricts the choices of producers. Take the owner of a furniture factory who understands the perils of climate change and wants to make the fossil fuel footprint of his operation as small as possible. He'd like to make changes in his supply chain and upgrades to lower the plant's emissions, but the up-front cost would be high enough to force him to modestly raise his prices. Unfortunately, without regulations to ensure that his competitors likewise improve their operations, he would lose considerable market share to the furniture producers who didn't clean up their acts and thus could keep their prices artificially low. Reluctantly, our furniture entrepreneur gives up on his plans. Because an insufficiently regulated industry creates a race to the bottom in terms of environmental citizenship, our factory owner lacks the freedom to do the right thing.

During a 2017 interview on the podcast *On the Media*, William Ruckelshaus, whom President Richard Nixon appointed as the founding administrator of the EPA, discussed the ways in which market competition pushes states into a race to the bottom. Even before he took the helm of the EPA, he had understood the need for government action rather than simply counting on voluntary choices. "I had seen it already in my home state of Indiana," he said in the interview, "that absent any government interference, not much was going to happen, no matter how bad the situation got. You couldn't rely on the individual causing the pollution to take steps themselves without being pushed by the government on a more or less common basis with their competitors."

Ruckelshaus also realized that sometimes environmental protection required government action at the federal level in order to establish a level playing field for the states. He recalled how other states made it more difficult for Indiana to adopt environmental regulations. "Having attempted to regulate industry from the state, doing that alone in one state was not going to do it because they would move someplace else. In fact, George Wallace, who was then the governor of Alabama, would take out ads in the Indianapolis newspapers saying, 'Come on down to Alabama, we need jobs. We don't care about the environment.'" Just as our federal government must sometimes level the playing field for the states, sometimes international trade agreements and treaties are needed to level the playing field for nations.

Not only can regulations prevent a race to the bottom, but they also can provide the freedom and guidance that will foster a race to the top. Research showing the decisive role government can play in spurring economic development led to a phone call in 2018 from the Royal Swedish Academy of Sciences to Paul Romer, a New York University economics professor who made his mark demonstrating that decisive government role. The call from the Swedish Academy informed Romer that he was the co-recipient of the Nobel Prize in economics. "A well-regulated market is an amazing machine for discovering new things," Romer told *The Washington Post* in an interview about his prize. "We've got to use our powers of collective decision-making to steer this innovation machine toward things that we can know that can benefit everybody," Romer added. "And there's just no way around the fact that the public has a role here."

Another economist, Mariana Mazzucato, has also risen to prominence through years of illuminating the relationship between government and innovation. A professor at University College London (UCL), she founded and directs UCL's Institute for Innovation and Public Purpose. Her 2013 book, *The Entrepreneurial State*, argues that the private sector does not have a monopoly on innovation. Government often takes the lead because it can invest in research without being hampered by the need to generate immediate profits, and it can take bigger risks than most private-sector actors are willing to shoulder. In a review of *The Entrepreneurial State*, Martin Wolf, chief economics commentator at the *Financial Times* and one of the world's leading observers of finance and the economy, writes, "Yes, innovation depends on bold entrepreneurship. But the entity that takes the boldest risks and achieves the biggest breakthroughs is not the private sector; it is the much-maligned state."

Mazzucato devotes an entire chapter to the representative example of Apple's iPhone. In a related article about the iPhone she writes, "Many of the revolutionary technologies that make it and similar devices 'smart' were funded by the U.S. government, such as the global positioning system (or GPS), the touchscreen display, and the voice-activated personal assistant, Siri." She adds, "Plus, of course, there's the development of the Internet, another government-funded venture, which enables the iPhone to be a valuable tool." In his review Wolf writes, "Apple put this together, brilliantly. But it was gathering the fruit of seven decades of state-supported innovation." The federal government even gave a $500,000 small-business grant to Apple in its vulnerable early stages. Note that despite being a huge beneficiary of the taxpayers' largesse, Apple has shown its gratitude by becoming a notorious tax dodger.

Mazzucato's vision for government's role extends beyond funding basic research and forming public-private partnerships, vital as those functions are. She promotes "mission-oriented" innovation, in which we, through our government, create policies and steer resources in ways that serve a public purpose. She wants us to direct much of our government's innovative power to what she and others sometimes call "grand challenges," such as climate change, health, and sustainable development. She thinks government should create and shape some new markets to nudge producers and consumers toward, as I would put it, a sustainable, just, and prosperous economy.

I'm impressed by the abundance of government innovation reported by Mazzucato, but I can't help but think of how much more abundant it would have been if government did not have to constantly fight anti-gub'mint attitudes. For instance, note this sentiment from Grover Norquist, the politically powerful president of Americans for Tax Reform: "The government creates wealth the way a tick creates blood." Such scorn coupled with the myth that all wealth and innovation flows from the private sector makes progress tough.

Among the prominent whingers about government are the predators, those hypocrites I mentioned in the introduction, the ones who pretend to hold a deep ideological opposition to regulation when in fact they only decry regulation that restrains their accumulation of wealth while quietly lobbying for regulation that advances their accumulation of wealth. Predictably, it is the public good and we ordinary citizens who are their prey. Their self-serving and sometimes corrupt manipulation of government has greatly damaged America's democratic institutions, resulting in many examples of the bad governing that gives credence to the campaign to discredit gub'mint. As I said in the introduction, this book focuses on the market's less scrutinized systemic problems rather than the much more publicized depredations of individuals. However, allow me a brief digression to remind us of why we need regulations to restrain those individuals, including the influential people who make up the predator class.

Say hello to Robert Murray, the chairman of the board and former CEO of Murray Energy Corporation, one of the largest coal-mining outfits in the United States. The abrasive Murray seems willing to express his hatred for regulations anytime, anyplace. Consider his press conference shortly after a catastrophic mine collapse at his Crandall Canyon mine, in Utah. At a time when the fate of six trapped miners was unknown, you'd have expected him to discuss the desperate rescue effort and to extend his sympathies to the gathering of worried families and friends. But Murray had different priorities. He belligerently rambled on about the importance of the coal industry and the scourge of bills introduced in Congress to fight global warming: "Every one of these global warming bills that has been introduced in Congress [will] eliminate the coal industry and will increase your electric rates four- to five-fold." Turns out the six men died.

Better regulations and better enforcement likely would have prevented the deaths of those miners. Murray kept claiming an earthquake caused the mine collapse, but experts disagreed, including the Mine Safety and Health Administration investigators, who blamed "unauthorized mining practices." Many pieces of evidence suggest Murray Energy was cutting regulatory corners. A federal government report indicated the mine was already unsafe three years before the Crandall Canyon disaster. More generally, Murray Energy racked up hundreds of violations at its various mines, including Crandall Canyon, just in the three years before the collapse. Small wonder that Murray and his managers took the fifth and refused to cooperate with congressional investigators.

The oddest bit of evidence against Murray emerged during his legal assault on John Oliver. Yes, that John Oliver, the topical comedian who hosts the HBO show *Last Week Tonight*. In 2017, Oliver did a long segment on the Trump administration's attempts to revive the coal industry. Oliver devoted considerable time to Murray, ranging from criticism of the Crandall Canyon mine collapse to calling Murray a "geriatric Dr. Evil," although, given that Murray had already threatened to sue if HBO aired the segment, Oliver put it this way: "I'm not going to say, for instance, that Bob Murray looks like a geriatric Dr. Evil, even though he clearly does." Oliver also singled out Murray's famed antipathy to regulations, citing the lawsuit he filed against a federal rule intended to reduce black lung disease. "If you even appear to be on the same side as black lung," said Oliver, "you're on the wrong fucking side."

As expected, Murray sued for defamation, but a few months later a West Virginia judge dismissed the case. The judge ruled that mockery like the "geriatric Dr. Evil" crack was protected as satire and Oliver's more substantial remarks, such as his criticism of Murray Energy's safety practices, were not defamatory because the reporting and sources were credible and the information appeared to be accurate. This finding of credibility by the court added to the pile of evidence regarding Murray Energy's lax observance of safety rules.

Murray apparently feels that government infringes on what he sees as his freedom to run his company however he wants.

I would argue that his dangerous mining practices constitute a much greater infringement, a violation of the rights of his employees to work in a safe environment. How much freedom did those six miners who died in Crandall Canyon have?

US

Reducing the harm done by greedy, ruthless individuals is not the only reason we need government regulation. There's also the matter of the rest of us. Even those of us with the best of intentions sometimes need regulating.

Let's say you want to buy a new refrigerator that will save you money on your electricity bill. Let's also say you want your children to live on a planet that isn't overheated. To serve both those purposes, you decide to buy an energy-efficient fridge. How do you choose one?

Well, you could try the "rational consumer" approach that lies at the heart of the market model. You should probably start by earning a college degree in mechanical engineering with a particular knowledge of thermodynamics. Next you'll want to carefully study detailed plans for scores of fridge models to see which appears to be the most energy-efficient. Then you'll need to actually purchase one each of all those models and test them to determine if they live up to those detailed plans. Earning a PhD in climate science also would be helpful so you could figure out how much impact an efficient fridge would actually have on global warming. And after finally making a well-informed decision about a fridge, you'd need to plunge into the next round of studies before buying an air conditioner, a car, or even something as seemingly simple as a bag of frozen shrimp.

Yes, it's tough being a rational consumer in this complicated day and age. No wonder economists like to explain how the market works using simple products that consumers know well. (Or think they know well—again, remember the shrimp.) I did a quick online search for rational choice theory to see what kinds of examples economists offered to illustrate the decisions consumers face. Here are the first three I found: apples versus oranges, a $3 bottle of water in a coffee shop versus a $1 bottle in an adjacent store, and pizza versus ice cream. I drew two conclusions from my little survey: economists must be a hungry bunch, and we consumers seldom possess enough information and expertise to make sufficiently informed choices regarding many products and services.

Fortunately, through government we can reduce this problem, as we have to some degree in the case of refrigerators and many other home appliances. For example, the federal government guides appliance shoppers with its Energy Star program, which certifies fridges and other appliances that meet its standards. The Energy Star program still depends on voluntary choices by consumers, but it does offer us a way to partly overcome our lack of information. Government can employ the engineers and climatologists so each of us doesn't have to try to become a specialist in all the fields that apply to our consumer choices, which include pretty much every field under the sun.

But voluntary programs like Energy Star have a serious drawback: they're voluntary. That opens the door to free riders. No doubt most ride free unaware or only slightly aware of the ripple effect of harm that will be caused by their choices. But some free riders—a minority of them, I hope—know full well the ramifications of their unsustainable behavior but nonetheless dump the responsibility of protecting the planet onto others. I suspect some of these uncaring free riders subscribe to the Ayn Randian notion of makers and takers and think of themselves as makers, yet their destructive actions in the public realm put them squarely in the taker category. They're like free-riding spouses who thoughtlessly toss their dirty clothes onto the bedroom floor and don't care that the bedroom manages to remain tolerably tidy only because their responsible partner cleans up the mess.

But why should the responsible folks have to pick up the dirty socks of Robert Murray, Ted Poe, and the rest of the mess makers? Regulations can compel these shirkers to do their fair share of the chores while also helping the unwitting free riders do the right thing. For example, in the 1980s, the federal government and some state governments began setting modest but mandatory standards for the manufacturers of home appliances. Being mandatory, the energy-saving improvements capture all the free riders. Without fanfare, these standards have made a huge impact. According to the American Council for an Energy-Efficient Economy (ACEEE), by 2013, refrigerators consumed only a third as much power as they did prior to the implementation of the regulations. The ACEEE reports that "altogether, efficiency standards adopted since 1987 are on track to save consumers and businesses more than $1 trillion through 2020 and to reduce global warming emissions by nearly 4 billion tons, or the equivalent annual emissions of 800 million automobiles."

Sure, some regulations flop. Maybe they're frustratingly vague. Maybe they're nitpicky beyond all reason. Maybe they're based on erroneous information. However, abundant studies have shown that regulations generally serve us well, and that's despite all the gub'mint-hating efforts to cripple them. For instance, the EPA estimates that the Clean Air Act saved 164,300 adult lives in 2010 and will save 237,000 annually by 2020, and at bargain prices. The Center for Progressive Reform studied many of the federal government's most significant standards and found that taken together their estimated benefits exceeded their estimated costs by roughly 8 to 1. And regulation has enjoyed this stunning success despite having one hand tied behind its back by laissez-faire mythology.

Perhaps most telling, in 2018, Trump's very own Office of Management and Budget quietly released its mandated annual report on the costs and benefits of federal regulations. Presumably the quiet release had something to do with the fact that the report belies the frequent griping of the president and his fellow Republicans about costly and burdensome regulations. A large majority of both the costs and benefits came from EPA rules, and the takeaway is that environmental regulations are great deals for America. From 2006 to 2016, compliance with EPA rules cost business between $54 billion and $65 billion while society reaped benefits between $196 billion and $706 billion. This takes me back to a Trump campaign rally in 2016 at which he declared to the crowd, "Excessive regulation costs our economy $2 trillion a year. Do you believe that?" "No" would be a good answer.

In recent years the enormous net benefits of regulation have seldom caused Republican administrations and Congresses to break stride in their race to eliminate regulations. Just ten days after taking office Trump signed a blunt-instrument executive order dictating that for every new regulation issued two existing ones must be eliminated. By December 2017, it appeared that the Trump administration had far exceeded the dictates of his order. In a White House photo op that month, Trump stood between a massive heap of paper meant to represent all the pesky regulations in the current *Federal Register* and a few short stacks of paper meant to represent the modest number of regulations that were in the 1960 *Federal Register*. A big red ribbon symbolizing red tape ran from one paper pile to the other and Trump cut it with a pair of giant scissors—gold, of course. In his remarks he said, "Instead of eliminating two old regulations for every one new regulation, we have eliminated twenty-two. Twenty-two. That's a big difference. We aimed for two for one, and in 2017, we hit twenty-two to one."

Almost immediately, various experts and fact-checking organizations discredited Trump's 22-to-1 claim, but that didn't stop the Trump administration from continuing to deploy the 22-to-1 line. Exaggeration aside, the administration actually has eliminated a fair number of rules and has greatly slowed the output of new rules. And Trump and friends are trying to eliminate, suspend, or delay plenty of other regulations, though they're finding that enterprise more difficult than expected. Congress, occasionally including a number of Republican members, has sometimes thwarted Trump, as when he tried to kill the popular Energy Star program. The courts also have slowed or stopped many of Trump's plans, such as yet another attempt to scuttle a regulation requiring a variety of appliances to be more energy-efficient. The judges presiding over these legal defeats have cited abundant problems with the administration's cases, such as skipping required processes, misunderstanding the laws, and simply getting the facts wrong. The upshot is that Trump's deregulatory team has not been nearly as successful as they say, but they have done considerable damage and they're intent on tossing more and more dirty socks onto the floor.

Happily, we don't need to keep picking up dirty socks, whether they're knowingly strewn about by predators or unintentionally dropped by well-meaning citizens. We can put our collective foot down and make sure more of those dirty socks end up in the hamper.

12

TRUE VALUE

"Absolutely!"

That was not the first word I heard from Yanni Gonzalez, but it was pretty close to the first. And I would hear it many more times from him in the course of our conversation, spoken in the same positive tone.

The thirty-something Gonzalez works as the communications director and programs coordinator for the Central California Asthma Collaborative (CCAC), a nonprofit in the San Joaquin Valley. "I always wanted to advocate for something good," he said. From an early age he was involved in various social efforts, endeavors that continued after he graduated from college. In 2016, his longtime interest in health communications took Gonzalez to CCAC.

The small staff of CCAC has taken on a big job. In the valley, one of every seven adults and one of every five kids suffers from asthma. That's roughly 600,000 adults and 250,000 kids. Chief among the causes is the foul air in this hotbed of industrial agriculture. Droughts, blowing dust, and triple-digit summer heat don't help, either. Asthma usually can be successfully managed with proper treatment, but in a cruel twist many of the residents of this asthma-intensive valley live in poverty and can't afford the necessary care. That's where CCAC comes in; the nonprofit provides its services for free.

Gonzalez spends a lot of his time on the road, driving to the little towns scattered amid the sprawling farms. Sometimes he attends health fairs, resource fairs, and the like to spread the word about the services CCAC offers. At other times he visits the homes of people who have come to CCAC in search of help.

Gonzalez tells me about a six-year-old girl with severe, uncontrolled asthma whose family he has been working with. I'll call her Lucia. Lucia had been struggling with this chronic disease for years. Her family couldn't afford to pay for all the needed medical care, and despite repeated efforts, her mom— I'll call her Alma—hadn't been able to find more than sporadic assistance. "This girl hardly got to play outside," said Gonzalez. "She would miss a lot of school days." He added, "If a child is missing school because of asthma, it often means the parent is missing work, so that's an economic hit." Gonzalez often sees firsthand the misery caused by unmanaged asthma. "The kids *hate* that they have asthma," he said. "Going to the doctor and ER constantly. They ask, 'Why me?' It's heartbreaking to hear that."

Not long ago CCAC scrounged up enough funding to expand into Lucia's area, and soon Gonzalez and her family got together. On his initial visit Gonzalez did an assessment of the family's house, looking for triggers that might exacerbate asthma symptoms. He also started teaching Alma how to manage Lucia's illness, such as regulating the indoor humidity, a crucial element in taming the disease. "Just knowing how humidity was affecting her child and her being able to control the humidity in her home definitely reduced the symptoms," said Gonzalez. He happily reports that Lucia is now able to run around outside and seldom misses school because of asthma attacks.

Gonzalez attributes much of CCAC's success to the close, long-term relationships the staffers form with families. "We try to be there for the family as much as we can and do as much for them as possible."

As much as possible. That phrase brings us to the all-too-familiar choke point for social service work: money. There's seldom enough to hire as many staffers as needed or to pay them what their labor is worth. Gonzalez is a busy guy. "It feels like two jobs some days," he said with a wistful laugh. He works hard, he has skills, he has a college degree, he has experience, and he overflows with enthusiasm, yet he earns only $19 an hour. And the only reason he earns even that much is CCAC's concerted effort to support its staff; comparable social service jobs in the area pay much less.

Gonzalez felt uncomfortable talking about money, but he agreed to discuss the matter when I explained that it was important to the point I was making about the allocation of resources. "One thing you should know about nonprofit workers is that we work very hard, wear many hats," he said. "And sometimes it seems like we're not getting paid enough." He added, "I love doing what I do, but I've got bills, you know, just like anyone else."

The same could be said for CCAC as a whole. "We'd like to help as many people as we can, but unfortunately our resources are limited," said Gonzalez. Kevin Hamilton, CCAC's CEO, echoed the sentiment: "We're not even touching 10 percent of the total need. If we had more money, I'd hire more people and I'd pay them better." Hamilton already spends half his time fundraising, but even if he spent all his time chasing dollars, their budget would still fall way short. Right now Gonzalez is the only CCAC person covering a big chunk of the valley. "I would love to have more support, have more people working alongside me," he said. I asked if he thinks there's a lot more demand for CCAC's help than they can provide. Inevitably, the word he chose for his answer was "Absolutely."

Not only does neglecting families with asthma constitute a moral failure; it also fails on practical grounds. "We believe strongly that we are saving a significant amount of money for the health care system in California," said Hamilton. He cited a study that estimates a health care savings of \$3.62 for every dollar CCAC spent. By building relationships and teaching families to manage their own asthma, CCAC prevents many expensive hospitalizations and ER visits. And this study didn't even include a price on intangibles, like the distress of a child struggling to breathe.

Given the double win society receives from allocating money to a place like CCAC, why aren't there a hundred more people like Gonzalez helping all those Lucias? And why does some hedge fund honcho make 25,000 times more a year than Gonzalez?

For the same reason our economy heedlessly degrades ecosystem services: the market. The market doesn't see the kind of work Gonzalez does, just as it doesn't see the kind of work wetlands and coral reefs do. The market doesn't send Wall Street investors to CCAC's door, just as it doesn't seek opportunities to invest in preserving wetlands and coral reefs. As we've seen in previous chapters, our market economy directs resources only to market goods and services and overlooks much of what matters most. As Gonzalez's example shows, this flaw in the market does damage well beyond the environmental realm. In many arenas our economic system falls short in its fundamental purpose: optimally allocating limited resources.

Focusing largely on jobs and investment, this chapter

will take a final look at why our economy often fails to put resources to their highest use and what we can do about it. And by looking at jobs from some unusual angles, we'll see some of the ways in which a sustainable economy would better our lives and our society (in addition to handling pesky market failures like climate change).

People who lean laissez-faire often view nonmarket jobs as a shabby sideshow and the market as the main event. Market disciples take the fact that nonmarket jobs typically pay modestly as a sign of lesser worth. As former Republican U.S. senator Rick Santorum said in a speech to the Detroit Economic Club during one of his runs for the presidency, "There is income inequality in America. There always has been and hopefully, and I do say that, there always will be. Why? Because people rise to different levels of success based on what they contribute to society and to the marketplace and that's as it should be." Wrong in so many ways, Santorum's remark brings to mind all sorts of public-private occupational pairings in which the levels of financial success seem way out of whack with the contribution to society. Consider firefighters versus currency speculators, EPA climate scientists versus oil company executives, kindergarten teachers versus tobacco company lobbyists, ad infinitum. The market flounders when it comes to recognizing true value.

The realization that the size of a person's bank account often does not reflect the worth of their work should discredit the subset of affluent people who fancy themselves the makers, the ones who despise paying some of their excessive riches in taxes to the so-called takers, a.k.a. the public. Deflating this inflated image of the wealthy is important because raising their taxes is one of the ways we can have more resources for the Gonzalezes of the world—or for whomever and whatever we collectively judge to be most worthy. Note that another subset of affluent people, including billionaires Bill Gates,

George Soros, and Tom Steyer, agrees that their taxes should be raised. There's even an organization of wealthy Americans, the Patriotic Millionaires, who lobby for higher taxes on the rich.

Yes, I said "taxes," and I'll say it again: taxes, taxes, taxes. Decades of anti-gub'mint crusading have rendered the word blasphemous, but we need to rescue the word and overcome our fear of saying it out loud. We need to think of paying taxes as spending money to buy a public service, just as we'd spend money to buy a new winter coat. If that coat is warm, well made, and reasonably priced, we'd consider its purchase a good deal. Likewise, if we do our jobs as citizens and fashion a government that serves our needs well and makes efficient use of our taxpayer dollars, we should consider that a good deal. For example, if we pay a bit more in taxes but get a livable climate in return, that's a bargain.

Raising taxes is only one of many tools for guiding more resources to the goods and services the market neglects. For example, the federal government could obviate some of the need for tax increases by redirecting its current harmful spending, such as the $20 billion plus in direct subsidies that it doles out annually to fossil fuel companies. (According to the International Monetary Fund, those subsidies soar to about $650 billion a year if we include indirect subsidies, such as the cost of environmental externalities.) We could enjoy an abundance of such savings if government simply followed the Hippocratic oath: "First, do no harm." Maybe our net taxes would go down, especially for the poor and middle class, even as we allocated more money to things that matter more.

Many of the public tasks that need doing wouldn't be done by government employees or rely on taxpayer funding, other than perhaps for getting started and for oversight. For example, regulations could direct more private-sector resources toward important nonmarket goods and services. Consider the step Californians took in 2018 when the state changed its building code to require that most new houses built after 2019 feature solar panels. This doesn't mean the state government is hiring roofers and solar installers; the market still does that. The new building code simply implements a collective decision that will steer additional workers and capital toward clean energy. The state estimates that requiring solar will add about $8,400 to the up-front cost of a home, but by directing that money to a public good, this requirement implicitly recognizes the value of the nonmarket services provided by a stable climate and clean air. Besides, as the executive director of the California Energy Commission noted, over the course of a thirty-year mortgage, solar will save an average homeowner about $19,000 on their energy bills. And the payback starts immediately; average homeowners will save about $80 a month on their energy bill while paying only $40 more a month on their mortgage.

As we gradually overcome our gub'mint phobia and direct more and more resources to nonmarket goods and services, we'll eventually find a desirable balance between private and public spending—a consciously chosen balance in which market and nonmarket goods compete. It will never be perfect. It's impossible to precisely calculate the relative merit of every good and service, private and public, given the previously noted impossibility of internalizing every cost and benefit. To create a reasonable balance we'll have to continually make adjustments.

Actions such as California's solar requirement certainly move sustainability forward, but to reallocate resources at the necessary scale we will also have to alter the DNA of our economic system. For example, we must throttle back on investing with short-term profit as the only goal. If a shift to sustainable investing on a significant scale seems unlikely to you, consider a recent decision made by the European Investment Bank (EIB), the world's largest international public financial institution. Under pressure from climate activists, in 2019 the bank's board agreed to stop loaning money to new fossil fuel development by 2021. This will largely turn off a spigot through which billions of dollars have been flowing to oil, gas, and coal projects.

The EIB decision marks a promising step in the right direction, but we'll have to dig deeper still, down into the systemic weeds. For instance, we might want to change who gets to create money.

Most people think that in America the federal government creates money. And that's true; the Treasury Department does print money, and the Federal Reserve sometimes expands the money supply. But that's only half the story. Or, to be more precise, about 5 to 10 percent of the story, because 90 to 95 percent of our money supply comes to life in the womb of private commercial banks. When one of these banks takes in a deposit—say $100—it can give birth to money in the real world by making loans based on that $100 deposit. Skipping over the mechanics of the process, the upshot is that the $100 deposit enables banks to conjure as much as $900 into digital existence and loan it to businesses and people in the real world—and the bank profits from the interest paid by the borrowers.

One might rightly question why the public should give private banks the lucrative privilege of creating most of our society's money and loaning it at interest, but that's not my concern here. One also might rightly worry, as do economists Herman Daly and Josh Farley in *Ecological Economics*, that this system will spur unsustainable growth as borrowers repay these conjured loans plus interest, constantly ratcheting up the money supply. But that's not my concern here, either. In the context of our discussion about steering more resources to nonmarket goods, I worry that private banks, quintessential creatures of the market that they are, will almost exclusively loan their vast, unearned riches to typical market enterprises.

Sustainability economists Peter Victor and Tim Jackson write about this problem in their paper "Towards a New, Green Economy": "There are a number of important implications of this debt-based money system. One of them is that the investments that are needed for the green economy must generally prove their creditworthiness on entirely commercial grounds and must compete for capital with all sorts of commercial investments. Potential green investments may have high social and environmental returns, but unless they also have attractive financial returns—an unlikely case—they typically won't be funded through commercial channels."

Victor and Jackson note that some enterprises attract commercial investment in part because they boost their bottom lines by ignoring the externalities they generate. As the authors write, "[Green investments] must compete with investments in dirty, extractive industries that degrade the environment, and operate through supply chains profitable only because they involve child labor." They add, "The social benefits of green investment are rarely factored into the commercial market. Neither are the social costs of unsustainable investment (including the huge cost of unrestrained speculative trading). Worse still, these social costs are often ultimately borne by the taxpayer rather than by the investors."

Victor and Jackson propose various solutions, as do other critics of the current process, but the details lie beyond the scope of this book. However, the thrust of these potential solutions entails giving the public, whether the federal government or local communities, more control over the money supply and therefore more say about how that money gets invested. But the benefits that come from allocating more resources to the public sector don't stop there. This more purposeful allocation could go a long way toward solving the seemingly endless struggle to create enough jobs while remaining within planetary boundaries.

GOOD JOB

One day I typed into Google "What is a job?" The first entry read: "A paid position of regular employment." The sample sentence showing how the word is used read—and I swear I'm not making this up—"Jobs are created in the private sector, not in Washington." If that's a quote from someone, it was not attributed. I'm going with the notion that Google's algorithm is a classic Silicon Valley libertarian. But that definition and sample sentence reveal how closely contemporary society associates "job" with earning money, especially in the marketplace. Jobs and the investments that create jobs are the main way in which we allocate the limited resources of labor and capital.

Perhaps Google's viewpoint about jobs is largely accurate today, but to shape the society we need for tomorrow we'll have to update that viewpoint. Or, more precisely, we'll need to go back in time to our premodern, tribal days for our understanding of the concept.

For most of the roughly three hundred thousand years that *Homo sapiens* have walked the earth, people worked at tasks important to their own individual survival and enjoyment and that of their families and tribes. Individually and collectively (unless they lived under the thumb of some all-powerful ruler) people decided what needed doing, and those were that society's jobs. No doubt they made plenty of bad decisions, but at least they were making a conscious effort to do meaningful work. That began changing with the advent of market economies. The market hasn't usurped all the power to determine which jobs get done, and many of the jobs the market creates are indeed meaningful ones that people happily choose to do. But by giving the market most of the power to allocate scarce resources, many of us end up laboring at jobs largely for the dollars. I've certainly done so more than once. Many of these jobs are relatively unimportant, and some even do more harm than good.

A market fundamentalist would say that if someone buys an item, then by definition that item and the jobs that produce that item are important, even if some elitist pinhead doesn't happen to think they're important; the market has spoken. But as we've established, the market often misspeaks. So once again we're going to need our judgment. In some cases we're going to have to collectively figure out the relative importance of competing goods and services, including nonmarket items. Should we put our resources into creating jobs that produce Joan Rivers Pavé Bluebird of Happiness Brooches or into jobs that produce electric vehicles? I realize this is a facile example, and such choices can get complicated, but I trust you get the idea. Though it's obviously impractical to weigh all the priorities and pass fully informed judgments on the relative worth of all the jobs in the world, we can make broad judgments that will provide broad guidance.

For example, I would judge that we should increase tax rates for the affluent and use some of that money to fund more people like Gonzalez, even if raising those taxes slowed the increase in what MIT economist David Autor labels "wealth work." According to Autor and a 2019 article in *The Atlantic* by Derek Thompson aptly called "The New Servant Class," occupations that largely cater to the desires of the well-off have proliferated in recent years. For instance, jobs as private household cooks have increased more than twice as fast as employment in general. Wealth work is mostly low-paid and contributes relatively little to society. I'm guessing most of these wealth workers would prefer jobs that are better paid and more rewarding. I'm also guessing that society would be better off if these workers were paid better and did more rewarding jobs. Sustainability economics could help make that happen by allocating more resources to pursuits that would make us happier both as consumers and as workers.

We should take happiness into account more often in our economic decisions. After all, what is an economy for if not to make us happy? Fortunately, happiness is starting to get a seat at the table. In that blockbuster 2019 UN report on biodiversity and ecosystem services I referred to earlier, the authors declare that humanity needs "a change to the definition of what a good quality of life entails—decoupling the idea of a good and meaningful life from ever-increasing material consumption." I think quality of life and happiness are pretty much the same thing.

A 2019 study in *Environmental Research Letters* explores the subject of human needs and wants and how they relate to sustainability. To establish the elements of *Homo sapiens* happiness, the study used the famous nine human needs devised by economist Manfred Max-Neef: subsistence, protection, affection, understanding, participation, leisure, creation, identity, and freedom. Then the researchers estimated the volume of carbon emissions associated with securing a reasonable amount of each of these needs.

The study finds that more abstract sources of happiness, such as participation (engagement with one's neighbors, attending church, belonging to associations, etc.), generate very little carbon. Meeting more materially based needs, such as ample food, health care, and decent housing, generates more emissions. In this category of material needs, however, when further growth takes us beyond a moderate threshold, the accompanying glut of stuff adds little to our happiness, but producing that surfeit of things pours copious amounts of carbon dioxide into our atmosphere. The upshot is that we can greatly reduce carbon emissions by dialing back growth while still enjoying ample stuff-related happiness through development. At the same time, with almost no increase in emissions, we could better meet our other, less material needs, which often get neglected in a market economy.

So maybe we should accelerate the creation of an economy that will help deliver these elements of happiness without hurting people or the planet. An economy that simply engorges the GDP just doesn't cut it. As Robert F. Kennedy liked to point out, the GDP measures everything "except that which makes life worthwhile."

Eventually, the transition to sustainability likely will result in an economy that doesn't generate as many jobs—at least not today's typical market jobs. Concern about a lack of jobs often causes progressives to join conservatives in worrying about the impacts of a sustainable economy. Sustainability economics provides possibilities that should soothe those worries. Let's start with the most straightforward solution: some people could work less and would be happy to do so.

Probably the most eminent economist to suggest that people would work less in the future was John Maynard Keynes. He fleshed out this idea in his famous essay "Economic Possibilities for Our Grandchildren," the first version of which he wrote in 1928. In "Economic Possibilities," Keynes imagined what the world's economy would look like a century later, in 2028, a date that we have nearly reached. He predicted that the ongoing revolution in technology and the subsequent economic growth would make satisfying our material needs so easy that most of us would need to work only about fifteen hours a week, giving us tons of leisure time.

Okay, let's all take a moment out of our sixty-hour work-weeks to guffaw. Working fifteen hours a day is more common than working fifteen hours a week. However, though his prediction was amusingly inaccurate, Keynes was right that technological and economic advances would move us inhabitants of the twenty-first century closer to the possibility of working fewer hours.

Trying to figure out why we twenty-first-century dwellers, even those in wealthy nations, have not taken advantage of the opportunity to work less has become a cottage industry. Explanations include that people continue to work long hours in order to buy more stuff than they need because more stuff confers social status; that people are addicted to stuff and shopping and can't quit excessive consuming; that many people in our vastly unequal society don't have enough and have no choice but to work long hours; and that people work more than necessary because people are hardwired to work and don't desire huge amounts of leisure time. I'd speculate that there's some truth in all those ideas. Still, I'd also speculate that a fair number of people would enjoy having more leisure time. (You can't see me, but my hand is raised.)

Vindicating Keynes somewhat, in wealthier societies even some middle-class people can indeed afford to work less and to some extent are choosing to do so. Several countries have instituted slightly shorter workweeks, many nations offer ample amounts of paid vacation, and many offer generous maternity and paternity leave. Workers in Germany, France, Norway, Denmark, and a dozen other European countries labor three hundred or four hundred hours fewer per year than Americans. Having less work available would create hardships for some people, especially in America due to our tattered social safety net, but solid safety nets like those in many European countries can lessen the problems that crop up when we turn down the throttle on work.

However, Keynes's fifteen-hour workweeks will have to wait a while longer. Most obviously, a lot of people will want to put in forty-hour weeks at better jobs for higher pay until they attain decent levels of income and wealth. Less obviously, our human tribe still has many important tasks that need doing. Keynes was looking at market goods when he predicted that by 2028 many of us in developed nations would have enough stuff and would be able to cut our work hours. But for the foreseeable future we could use a lot more nonmarket goods and services as well as more of the kinds of market goods that will speed the transition to sustainability. We need all hands on deck to clean up polluted rivers, care for our elderly, teach music, manufacture wind turbines, staff after-school programs, grow food sustainably, start green businesses, build affordable housing, and get the lead out of our drinking water. Many are service jobs that don't involve making a lot of stuff and therefore don't push our environmental limits. Many involve making stuff but stuff that we need for a sustainable future. And many market and nonmarket jobs alike can be done using sustainable methods. This is how we continue developing without resorting to unsustainable growth while also producing plenty of jobs.

GO FAST, GO BIG

All this talk of important, sustainable, decently paid jobs drops us squarely into the lap of the Green New Deal (GND). The basic idea of the GND hovered on the fringes of the political economy for years, occasionally dipping its toe into the mainstream. However, within a few weeks after the 2018 midterm elections, it plunged into the current, propelled by the big influx of Democrats into Congress. Within a few months many Americans had heard of the GND and scores of congressional Dems had signed up to support it, at least in broad strokes.

What did they sign up for? Well, at this point that's not entirely clear because the GND is a work in progress. Its framers want to gather input from all sorts of people to help guide its development. At the time of this writing the GND is less a detailed blueprint than an aspirational outline. Still, the essential character of the GND became evident in 2019, especially after its most visible advocates, Representative Alexandria Ocasio-Cortez (D-NY) and Senator Ed Markey (D-MA), released the text of their proposed nonbinding congressional resolution. (A nonbinding resolution of this kind is meant to demonstrate intent and provide guidance on an issue; it is not an actual bill that would have the force of law. In a tweet, Ocasio-Cortez advised thinking of the resolution as a "request for proposals.")

Despite the resolution's brevity, it reveals breathtaking ambition. Though largely built on climate action, the goals range from classic environmental targets, like clean air and clean water, to justice and equity aims, such as providing economic security for all Americans and easing the oppression of marginalized communities. The GND's backers intend to rapidly mobilize around these aims and make significant progress within the next decade. The initial plan proposes to attain net-zero carbon emissions by 2050 while progressing toward all those social goals. Like I said, ambitious.

Naturally, the GND has sparked a ferocious debate among politicians and pundits. Some are arguing over the big picture. Is the GND a government power grab that will bleed freedom from America's veins? Or is it a democratic reform that will put America on a path to sustainable prosperity? Others are arguing over details. Will the GND count nuclear power as clean energy? Will it create a sufficient number of green jobs in marginalized communities? No doubt politicians and pundits will produce terabytes of conflicting opinions regarding the merits of the GND for as long as it remains in the lime-light.

Predictably, the scorched-earth assaults on the GND are coming from conservative politicians and pundits. However, though agreeing that something must be done about climate change, numerous centrist Democratic politicians and pundits also have qualms about the GND and are trying to rein it in. Typically, they view the GND as politically unrealistic, as striving to accomplish too much too soon, and as including too many items from a progressive wish list. Maybe. It's a debate worth having. But the centrist preference for slow, incremental change is biophysically unrealistic because it overlooks the fearsome scientific realities of climate change and other environmental crises. It's also socially unrealistic because it fails to summon the urgency needed to address long-standing inequities.

When it comes to global warming, after decades of obstruction by Republicans and dithering by centrist Democrats, we now have no choice but to go fast and go big if we want to soften the bruising climate blow we're sure to take and lower the risk of the knockout punch we might be facing. If not the GND, then some other muscular plan is needed. By all means, centrists should feel free to criticize the GND—after all, the people formulating it are seeking input—but it should be constructive criticism that helps us go fast and go big because global warming won't wait for a more politically convenient time.

When we look beyond the politicians and pundits, a funny thing happens; disapproval of the GND shrinks dramatically. Take a poll conducted by the Yale Program on Climate Change Communication and the George Mason University Center for Climate Change Communication. This survey found that 81 percent of registered voters in the United States support the GND. When the surveyors sliced and diced the numbers, they found bipartisan approval: 92 percent of Democrats, 88 percent of independents, and 64 percent of Republicans expressed support.

Does this poll presage the rapid passage of GND legislation? According to my survey of the one registered voter who wrote this book, the answer is "no." Sure, most Americans like the GND ideas listed in the poll, such as investing in clean energy research, upgrading the nation's transportation infrastructure, and providing training for jobs in the new green economy. But at the time of the survey, the partisan attacks had not yet kicked into high gear. Republican politicians and pundits almost uniformly despise the GND, and as it gains attention they are escalating their war against it.

Consider the fundraising email sent out in 2019 by the honorary chairman of Protect American Values, a conservative super PAC (political action committee). That chairman is none other than Jack Abramoff, the poster boy of lobbyist corruption in the early 2000s. Convicted of mail fraud, tax evasion, and conspiracy to bribe public officials, Abramoff spent several years in prison. His notoriety reached such heights that there was a 2010 film about him called *Casino Jack*, starring Kevin Spacey. Abramoff spiced his anti-GND email with nuanced, insightful analysis, such as "Alexandria Ocasio-Cortez wasn't even in diapers when I started fighting the Left" and "These socialists are so confused, and their messaging is getting worse by the minute. They seem to think that cows and airplanes are a threat to our future. Sounds to me like they have been infected by Mad Cow Disease."

Such attacks have been mounting and they've been rapidly lowering the popularity of the GND among Republican voters. A study published in *Nature Climate Change* showed a plunge in support for the GND among registered GOP voters just in the four months from early December 2018 to early April 2019. Most striking was what the researchers termed the

"Fox News effect." Among Republican voters who watched Fox News once a week or less, approval of the GND dropped from 71 percent to 56 percent. Among Republican voters who watched Fox News more than once a week, support plummeted from 54 percent to 22 percent.

Even if a significant percentage of Republican voters still support the GND after it passes through the right-wing media meat grinder, the GOP-controlled Senate and the Trump administration would stymie any GND-related legislation. No genuine GND package has a chance of becoming official federal policy until at least January 2021, at which time the results of the 2020 elections might have altered the balance of power in D.C.

Many of the early salvos aimed at the Green New Deal consist of overwrought warnings about rampant socialism and the GND's supposed intent to propagate Soviet-style authoritarianism. Yes, the GND includes some traditional social democratic goals, such as promoting a living wage and high-quality universal health care, goals that are hardly limited to social democrats. But the Markey–Ocasio-Cortez resolution focuses a great deal on greening the economy, especially manufacturing, via government collaboration with the private sector—not by government taking over the private sector. Using the allocation of research dollars, subsidies to sustainable industries, regulations, and many other tools, the government can cultivate a sustainable economy without going Soviet and dictating how many blue and how many green widgets to produce.

We're already starting to experience the fruits of government sustainability policy at the state and local levels. Consider the boom in clean energy jobs in the rural Midwest—not exactly the typical magnet for green initiatives. A 2018 report from the Natural Resources Defense Council found that in the rural reaches of the dozen states studied, fossil fuel

employment was fading while clean energy jobs were boom-
ing due to a blend of government guidance and market forces.
Already clean energy jobs far outnumber fossil fuel jobs in
the region, and that gap is widening as employment in clean
energy occupations grows at an annual rate of some 5 or 6
percent. And this is a region whose economy-wide job growth
is crawling along at under 1 percent. A study from University
College London found that the rapid growth of green jobs is a
national trend in America. From 2013 to 2016 about 1.5 million new workers entered sustainability-related occupations,
an increase of about 20 percent.

While some conservative leaders have been flinging child-
ish insults and spreading conspiracy theories (the GND wants
to take away our cars and hamburgers!), other conservatives
have been attacking the GND as too expensive. This raises
valid questions, questions that other observers across the
political spectrum have also been asking: How much would
the GND cost and how would we pay for it? The GND brain
trust has long known cost is a vital issue and they offer a number of answers but none that has put the issue to rest. However, several credible studies estimate surprisingly low costs
for fast-and-big climate efforts. University of Massachusetts
Amherst economist Robert Pollin, for example, calculates
that the United States could meet the stringent IPCC goals
for emission reductions by spending about $600 billion a
year from now until 2050—a mere 2 percent of our projected
GDP. Most of the investments would be made by the private sector, guided by both government regulations and government assistance—sticks and carrots, as Pollin puts it. He
emphasizes the fact that over time these investments would
pay for themselves due to the long-term savings from clean
energy.

Though the public knows little about the costs of addressing climate change, the general idea of costs significantly

influences how people feel about GND goals, as indicated by a 2019 survey conducted for the think tank Data for Progress. For example, without a price tag attached, a sizable majority of respondents favored the GND's proposed mandate requiring 100 percent of America's electricity to be generated by renewable energy by 2050. But support dropped 6 percent when surveyors stuck a hypothetical price tag of $50 billion on the shift to renewables.

Concern about the costs of climate action brings us back to the fundamental thinking behind the efforts to determine an SCC; when making decisions about climate policy, we must consider the benefits of having a livable climate as well as the price of securing a livable climate. As I pointed out earlier in the book, the costs could be minimal and quickly turn into net benefits because a sustainable economy might rapidly generate greater prosperity than our current economy, even according to the cramped standards of the GDP. But I also pointed out the uncertainty surrounding the transition to a sustainable economy and the possibility that in the short term it may be costly in GDP terms. We just don't know. But we do know that business-as-usual greenhouse gas emissions would eventually cost us more—both in GDP terms and in terms of social costs—than what it would cost to curb emissions. Probably far, far more. Don't forget escalating wildfires, the dismal theorem, declines in agriculture, the Centre for the Study of Existential Risk, climate refugees, tipping points, chronic drought, mega-catastrophes, stunted economic growth rates, lethal heat waves, damages estimated at $200 trillion, and the fat tail risk of the collapse of civilization. Let's stick that information in the prompt that introduces a survey and then see how people feel about spending a relatively few bucks on Green New Deal goals.

Costs may be the leading concern for some people, but for laissez-faire ideologues another, more elemental characteristic of the GND likely inflames them above all else: the GND is a big, bold gub'mint initiative. "Dealing with climate change in the way that we need to is not just about passing a suite of policies that will transform our society to both end the causes of climate change and prepare society for the climate change that is already baked in," said Evan Weber, the cofounder of the Sunrise Movement, a youth organization that is a principal backer of the GND. "It's also changing our conception of what government is and who it's for." It is especially the implication of Weber's last sentence that triggers dread among those who fear gub'mint.

Much of the government engagement envisioned by the GND would be proactive, which touches a sensitive laissez-faire nerve. Not only does the GND aim to correct market failures, but it also aims to occasionally mobilize government to take the initiative in improving the economy. The GND would enable us to sometimes collectively decide what kinds of jobs need doing instead of letting a half-blind market make all the decisions for us. This smacks of "industrial policy," a strategy for economic development in which the government boosts certain sectors, especially in manufacturing, that it deems worthy. Many nations have successfully engaged in industrial policy, but in the United States, conservatives have largely driven the idea to the margins, deriding industrial policy as gub'mint "picking winners and losers" instead of leaving that to the market.

However, in the last few years, industrial policy (though typically not the term itself) has begun appearing in the mainstream, notably in the proposed policies of some of the Democratic candidates for president in the 2020 race. Numerous studies have shown that many aspects of industrial policy appeal to Americans. A 2019 YouGov Blue/Data for Progress survey asked registered voters what they thought of "economic nationalism for climate change," defined as something akin to industrial policy focused on government efforts to ramp up manufacturing in the clean energy sector. Most people thought it sounded good, with 53 percent supporting the idea and only 30 percent opposing. Perhaps the GND and its ilk have come along at a propitious time. Maybe knee-jerk resistance to all things gub'mint is softening and we will be able to shape an economy that hits the sweet spot, letting the market do what it does best, empowering the government to do what it does best, and enabling the market and government to work in concert when that is best.

What will the GND grow up to be? Will it ever become the law of the land? Would its implementation propel us toward the systemic transformation our economy needs? Will its trailblazing open the way for other aggressive sustainability plans? Intriguing questions that may be answered in the next few years. But in and of itself, the initial enthusiasm for the embryonic GND signals exciting progress, a potential for going fast and going big. It reveals a hunger for energetic collective action, a potential antidote to some of the anti-gub'mint fever that has sickened our political economy for decades. A lot of people seem ready, even eager, to use democracy to create a sustainable, just, and prosperous economy.

YES, WE CAN. BUT WILL WE?

That growing eagerness to wield democracy may be the key to creating the political will we'll need. Remember the findings of the research paper I cited in the previous chapter: political will requires committed support among key decision-makers for a particular policy solution to a particular problem. And, as noted, building that support among our political leaders usually requires outside pressure. The greatest pressure may come from us citizens, but it also may come from a self-serving subset of wealthy political donors and other members of the power elite. If we want to end up with an economy that serves us and our values, we're going to have to put our shoulder into political engagement and make sure we, the people, apply the most pressure.

But that's just the first step. If we manage to apply enough pressure to overcome the powers that be, then we must take the next step and use the resulting political will to push for the right actions, which is far from a given. After all, many American citizens and political leaders are still fighting against government action to curb global warming. More broadly, many are still resistant to almost any gub'mint action, which plays right into the hands of the powers that be. Under such circumstances, how can we pull together enough collective action and aim it at establishing a sustainable economy?

It won't be easy. Still, I think there's room for hope and even some optimism. A number of the pro-sustainability perspectives explored in this book seem to be moving from the margin and into the mainstream, though at this point they're seeping in more than pouring in. But that could change quickly in light of our accelerating environmental and social crises.

Consider the growing realization that supposedly remote and amorphous threats like climate change are in fact immediate and concrete. Some opponents of climate action try to portray the issue as removed from our everyday lives, but that notion appears more foolish by the day as more and more people suffer from climate-magnified floods, droughts, diseases, famines, fires, and hurricanes. Perhaps the time is ripe for channeling the apprehension roused by these crises into the solutions offered by sustainability economics. Maybe enough people are open to the kinds of ideas discussed in this book, heartening ideas that small but increasing numbers of economists, scientists, politicians, activists, and others are advocating.

Think back to our discussion of externalities. If more of us come to understand their ubiquity and their terrible social costs, more of us would be motivated to support systemic changes that reduce externalities. I don't think it's naive to believe that most people would favor better regulation of shrimp production if they knew that some producers used slaves.

A fuller grasp of externalities would lead to a fuller grasp of our market system's limitations, including such crucial matters as its inability to allocate resources to public goods. Greater awareness of those limitations would help take the market down from its pedestal. That would help people to stop worshipping the market and see it for what it is: a tool. The market can be an effective, even brilliant, tool when it is used for the right job, but relying on the market to deal with something like global warming is like trying to saw wood with a hammer. Demythologizing the market would also make it easier to separate it from some of the values that are often mistakenly associated with it, such as freedom, responsibility, and self-reliance. People who see the market for what it is could also see that misapplying the market actually undercuts these values, whereas sustainability economics enhances them.

I think a lot of citizens could be drawn to sustainability economics because it gives us the opportunity to be free, responsible, and self-reliant by exercising more control over the economy instead of surrendering so much autonomy to the market. As we start along this path, we'll wobble a bit; our decision-making muscles have atrophied, and we'll need time to strengthen them. But we'll get strong again, and soon we'll grow beyond being mere consumers and become forceful citizens.

Putting the market in its proper place would make more space for government. But to take full advantage of that space, government will have to overcome decades of demonization, and it will have to get genuinely better. This presents us with a catch-22: the demonization makes it hard to motivate us citizens to do our part in bettering government, but until government does better and demonstrates its value, we'll struggle to overcome the demonization.

I see no single grand solution, but I think a number of things can get the snowball rolling, and eventually it could gather enough momentum to continue on its own. For example, federal, state, and local governments already do many beneficial things, which we should acknowledge and appreciate. Many people have lost sight of even the most obvious government largesse. A 2012 Cornell University study notes that 53 percent of students receiving government loans and 40 percent of people on Medicare declared that they weren't getting any government benefits. We should also bear in mind the existential necessity of government; one of the principles of sustainability economics tells us that collective action is an indispensable element in addressing many of the nonmarket issues we face, such as climate change. Perhaps most important, we should own up to our responsibilities as citizens. Sure, it's entirely valid to lambast the too-numerous lamentable elected officials who befoul government, but, after all, they wouldn't be our elected officials if we hadn't elected them.

But our responsibilities go far beyond elections. Democracy will generate governments that are only as good as the effort that we citizens put into them. Elevating our effort is partly a quantitative matter; we need to spend significant time learning about government, engaging with elected officials, talking with fellow citizens, attending civic meetings, and maybe even running for office. Elevating our effort is also a qualitative matter. We need to tune out the ranting blowhards and conspiracy theorists. We need to question and hold to account the leaders of our own political factions as well as those of other factions. We need to be discerning about the media we consume and stop dismissing as "fake news" everything that doesn't support our existing positions. We need to care about the public good as well as our private good—and to realize that there's a great deal of overlap between the two. I know this sounds idealistic, and it is. None of us will do all this perfectly. But if enough of us exercise our citizenship well enough, we can nurture government that is good enough—and maybe, eventually, even better than good enough.

In this era of rancorous partisanship, anti-gub'mint fervor, and corporate influence, the notion that we can establish good government likely strikes many of you as unrealistic. You probably doubt that a tsunami of citizens will rise to the occasion. I share your doubts. Though I hope that tens of millions of disengaged Americans will suddenly throw themselves into governing and do so in quest of a sustainable, just, and prosperous society, I realize this may not happen, especially not soon enough for us to deal with urgent issues like global warming. My fallback hope is that a critical mass of citizens will embrace the ideas of sustainability economics and compel decision-makers to enact some of these ideas, thus creating the necessary political will to make some key systemic changes. Then the systemic changes might carry us at least partway up the path to a sustainable economy.

Still too idealistic? Perhaps, but I don't think so. The main reason I believe a critical mass of citizens will make it up that path is that it leads to such an appealing destination. Sustainability economics offers so much more than staving off environmental and social catastrophe (though that should be more than enough reason for us to follow the path). Most of the economic steps we must take to deal with climate change, environmental degradation in general, and social inequities are also steps we should take anyway to foster a more felicitous society. Building a sustainable economy is not like grudgingly eating rotten lemons to forestall scurvy. It's more like cheerfully feasting on an array of delicious dishes that will enhance your health because they're as nutritious as they are tasty.

So what are the odds that we will do what needs doing? Honestly, I don't know. I think it's better if I leave that answer to you. After all, you hold the power in your hands. However, I will say again what I said at the start of this book: though I'm not sure if we will create a sustainable, just, and prosperous economy, I am sure that we *can*. We just have to summon the political will—and that will happen if enough of us assert our public will.

ACKNOWLEDGMENTS

Climate change and economics are complicated enough when approached separately. Put them together and . . . well, let's just say that without guidance from many knowledgeable, dedicated, and patient people I might never have made it through this book alive. I'm grateful to all of them and sorry I can't mention them all.

Early on, several key sources helped me find my way to the heart of my topic, among them Ruth Greenspan Bell of the Wilson Center, Eban Goodstein of Bard College, Karl Hausker of the World Resources Institute, and Kristen Sheeran, who advises Oregon's governor on energy and climate change.

Once I'd found my core subject, I expanded the circle of informed people from whom I sought information. I learned a great deal from a number of iconoclastic scholars whom I think of as sustainability economists, including James K. Boyce of the University of Massachusetts Amherst, Jon D. Erickson of the University of Vermont, Josh Farley of the University of Vermont, Peter Howard of the Institute for Policy Integrity at New York University, and the late Frank Ackerman, long a leader among the iconoclasts.

Numerous other estimable economists representing all sorts of viewpoints also generously gave me their time: Dean Baker of the Center for Economic and Policy Research, Marshall Burke of Stanford University, Elliott Campbell of the state of Maryland, Michael Hanemann of Arizona State University, happily retired Chris Hope recently of the University of Cambridge, David Kreutzer of the Heritage Foundation, Richard Tol of the University of Sussex, Alan Viard of the American Enterprise Institute, and Benjamin Zycher of the American Enterprise Institute.

Shifting from the dismal science, I'd also like to express my appreciation to the scientists, government workers, historians, business owners, lawyers, and other non-economists who provided me with so much knowledge and insight. They include John Charles of the Cascade Policy Institute; Sean Donahue of Donahue, Goldberg, Weaver & Littleton; Yanni Gonzalez of the Central California Asthma Collaborative; Kevin Hamilton of the Central California Asthma Collaborative; Dina Kruger of Kruger Environmental Strategies; Nancy MacLean of Duke University; Thomas McGarity of the University of Texas School of Law; Phil Mote of Oregon State University; Peter Muennig of Columbia University; Julia Olson of Our Children's Trust; Jim Petitto of Petitto Mine Equipment; David Pinsky of Greenpeace; Richard Revesz of the New York University School of Law; and Andrea Rodgers of Our Children's Trust.

If they awarded medals for being patient with tardy authors, Andrew Miller, my editor, would surely have gotten one for the exceptional forbearance he displayed as I dragged my book across the finish line. I'm also deeply grateful to my other editor, Maris Dyer, who worked with me day to day and helped me steer through the shoals of revision to arrive at the best book possible. Thanks, too, to Mary Beth Constant, my copy editor with the fitting last name, and the rest of the Anchor team for turning a manuscript into a book. And without the guidance of Robin Straus, my agent, I never would have gotten started with this book in the first place.

Finally, I want to thank my family and friends. They supported me in so many ways, from reading parts of the manuscript to making sure I occasionally left the keyboard and had some fun. A special thanks to the three family members who live with me or close by: Mary, my wife; Sarah, my daughter; and David, my son-in-law. They kept me more or less sane.

BIBLIOGRAPHY

Ackerman, Frank. "Richard Tol on Climate Policy: A Critical View of an Overview." Synapse Energy Economics, July 21, 2014.

———. *Worst-Case Economics: Extreme Events in Climate and Finance.* London: Anthem Press, 2017.

Ackerman, Frank, and Lisa Heinzerling. *Priceless: On Knowing the Price of Everything and the Value of Nothing.* New York: New Press, 2004.

Ackerman, Frank, and Elizabeth Stanton. "Climate Risks and Carbon Prices: Revising the Social Cost of Carbon." *Economics: The Open-Access, Open-Assessment E-Journal* 6 (May 13, 2012).

Ault, Toby R., et al. "Relative Impacts of Mitigation, Temperature, and Precipitation on 21st-Century Megadrought Risk in the American Southwest." *Science Advances* 2, no. 10 (2016).

Bevis, Michael, et al. "Accelerating Changes in Ice Mass Within Greenland, and the Ice Sheet's Sensitivity to Atmospheric Forcing." *Proceedings of the National Academy of Sciences* 116, no. 6 (February 5, 2019): 1934–39.

Bezdek, Roger. *The Social Costs of Carbon? No, the Social Benefits of Carbon.* Management Information Services, January 2014.

Boyce, James K. *The Case for Carbon Dividends.* Medford, MA: Polity Press, 2019.

———. *Economics for People and the Planet: Inequality in the Era of Climate Change.* London: Anthem, 2019.

Brennan, Pat. "Narwhal Recruits Track Melting Arctic Ice." *Global Climate Change*, NASA, October 26, 2017.

Brondizio, Eduardo, et al., eds. *Global Assessment Report on Biodiversity and Ecosystem Services.* Intergovernmental Science-Policy Platform on Biodiversity and Ecosystem Services, 2019.

Burke, Marshall, Solomon M. Hsiang, and Edward Miguel. "Global Non-Linear Effect of Temperature on Economic Production." *Nature* 527 (2015): 235–39.

Center for Progressive Reform, https:www.progressivereform .org/.

Centre for the Study of Existential Risk, https:www.cser.ac.uk/.

Costanza, Robert, et al. "Changes in the Global Value of Ecosystem Services." *Global Environmental Change* 26 (May 2014): 152–58.

Daly, Herman, and Joshua Farley. *Ecological Economics: Principles and Applications.* Washington, D.C.: Island Press, 2011.

Daniel, Kent D., Robert B. Litterman, and Gernot Wagner. "Declining CO_2 Price Paths." *Proceedings of the National Academy of Sciences* 116, no. 42 (October 15, 2019): 20886–91.

Dietz, Robert, and Daniel W. O'Neill. *Enough Is Enough: Building a Sustainable Economy in a World of Finite Resources.* San Francisco: Berrett-Koehler, 2013.

Drijfhout, Sybren, et al. "Catalogue of Abrupt Shifts in Intergovernmental Panel on Climate Change Climate Models." *Proceedings of the National Academy of Sciences* 112, no. 43 (October 27, 2015): 5777–86.

Environmental Protection Agency Office of Air and Radiation. "Benefits and Costs of the Clean Air Act from 1990 to 2020." April 2011.

Farley, Joshua, et al. "Extending Market Allocation to Ecosystem Services: Moral and Practical Implications on a Full and Unequal Planet." *Ecological Economics* 117, no. 215 (2015): 244–52.

Farley, Joshua, and Robert Costanza. "Envisioning Shared Goals for Humanity: A Detailed, Shared Vision of a Sustainable and Desirable USA in 2100." *Ecological Economics* 43 (2002): 245–59.

Friedman, Milton, and Rose Friedman. *Free to Choose.* Boston: Houghton Mifflin Harcourt, 1980.

Galbraith, James K. *The Predator State: How Conservatives Abandoned the Free Market and Why Liberals Should Too*. New York: Free Press, 2008.

Greenstone, Michael, Elizabeth Kopits, and Ann Wolverton. "Developing a Social Cost of Carbon for U.S. Regulatory Analysis: A Methodology and Interpretation." *Review of Environmental Economics and Policy* 7, no. 1 (Winter 2013): 23–46.

Gustafson, Abel, et al. "The Development of Partisan Polarization over the Green New Deal." *Nature Climate Change* 9 (2019): 940–44.

Harvey, Hal, Robbie Orvis, and Jeffrey Rissman. *Designing Climate Solutions: A Policy Guide for Low-Carbon Energy*. Washington, D.C.: Island Press, 2018.

Hausfather, Zeke. "Mapped: The World's Largest CO_2 Importers and Exporters." *CarbonBrief*, July 5, 2017.

Hawken, Paul, ed. *Drawdown: The Most Comprehensive Plan Ever Proposed to Reverse Global Warming*. New York: Penguin, 2017.

Hope, Chris W. "The Social Cost of CO_2 from the PAGE09 Model." *Economics: The Open-Access, Open-Assessment E-Journal* (September 15, 2011).

Howard, Peter. "Flammable Planet: Wildfires and the Social Cost of Carbon." Cost of Carbon Pollution project, Institute for Policy Integrity, September 2014.

———. "Omitted Damages: What's Missing from the Social

Cost of Carbon." Cost of Carbon Pollution project, Institute for Policy Integrity, March 13, 2014.

Howard, Peter, and Jason Schwartz. "Foreign Action, Domestic Windfall: The U.S. Economy Stands to Gain Trillions from Foreign Climate Action." Institute for Policy Integrity, November 2015.

———. "Think Global: International Reciprocity as Justification for a Global Social Cost of Carbon." *Columbia Journal of Environmental Law* 42, no. S (2019).

Hsiang, Solomon, and Trevor Houser. "Don't Let Puerto Rico Fall into an Economic Abyss." *New York Times*, September 29, 2017.

Idso, Craig D. "The Positive Externalities of Carbon Dioxide: Estimating the Monetary Benefits of Rising Atmospheric CO_2 Concentrations on Global Food Production." Center for the Study of Carbon Dioxide and Global Change, October 21, 2013.

Interagency Working Group on Social Cost of Carbon. "Technical Support Document: Social Cost of Carbon for Regulatory Impact Analysis." February 2010.

Jackson, Tim. *Prosperity Without Growth: Economics for a Finite Planet.* London: Earthscan, 2009.

Jackson, Tim, and Peter A. Victor. "Towards a New, Green Economy." Next System Project, October 2016.

Johnson, Laurie T., and Chris Hope. "The Social Cost of Carbon in U.S. Regulatory Impact Analyses: An Introduction and

Critique." *Journal of Environmental Studies and Sciences* 2, no. 3 (September 2012): 205–21.

Klein, Naomi. *This Changes Everything: Capitalism vs. the Climate.* New York: Simon & Schuster, 2014.

Larsen, John, et al., of the Rhodium Group. "Energy and Environmental Implications of a Carbon Tax in the United States." Columbia/SIPA Center on Global Energy Policy, July 17, 2018.

Lenton, Timothy M., et al. "Climate Tipping Points—Too Risky to Bet Against." *Nature*, November 27, 2019.

———. "Tipping Elements in the Earth's Climate System." *Proceedings of the National Academy of Sciences* 105, no. 6 (February 12, 2008): 1786–93.

Lustgarten, Abrahm. "Fuel to the Fire: How a U.S. Law Intended to Reduce Dependence on Fossil Fuels Has Unleashed an Environmental Disaster in Indonesia." ProPublica and *New York Times*, November 20, 2018.

Mabus, Ray. "Don't Ignore Military Advice on Climate Change, Mr. President." *Military Times*, March 6, 2019.

MacLean, Nancy. *Democracy in Chains: The Deep History of the Radical Right's Stealth Plan for America.* New York: Viking, 2017.

Mankiw, N. Gregory. *Principles of Economics.* Boston: Cengage, 2018.

Mason, Margie, et al. "Global Supermarkets Selling Shrimp Peeled by Slaves." Associated Press, December 14, 2015.

Mazzucato, Mariana. *The Entrepreneurial State: Debunking Public vs. Private Sector Myths.* New York: PublicAffairs, 2015.

———. *The Value of Everything: Making and Taking in the Global Economy.* New York: PublicAffairs, 2018.

Moore, Frances, and Delavane Diaz. "Temperature Impacts on Economic Growth Warrant Stringent Mitigation Policy." *Nature Climate Change* 5 (2015): 127–31.

Muennig, Peter. "The Social Costs of Childhood Lead Exposure in the Post-Lead Regulation Era." *Archives of Pediatric and Adolescent Medicine* 163, no. 9 (September 2009): 844–49.

Nordhaus, William. *The Climate Casino: Risk, Uncertainty, and Economics for a Warming World.* New Haven, CT: Yale University Press, 2013.

———. "Revisiting the Social Cost of Carbon." *Proceedings of the National Academy of Sciences* 114, no. 7 (February 14, 2017): 1518–23.

O'Neill, Daniel W., et al. "A Good Life for All Within Planetary Boundaries." *Nature Sustainability* 88, no. 95 (2018).

Ostrom, Elinor. *Governing the Commons: The Evolution of Institutions for Collective Action.* Cambridge, UK: Cambridge University Press, 1990.

Our Children's Trust, https://www.ourchildrenstrust.org/.

Pell, M. B., and Joshua Schneyer. "Reuters Finds 3,810 U.S. Areas with Lead Poisoning Double Flint's." Reuters, November 14, 2017.

Pickett, Kate, and Richard Wilkinson. *The Spirit Level: Why Greater Equality Makes Societies Stronger.* London: Penguin UK, 2010.

Pollin, Robert. "Global Green Growth for Human Development." United Nations Development Programme, 2016.

Post, Lori Ann, et al. "Using Public Will to Secure Political Will." In *Governance Reform Under Real World Conditions*, edited by Sina Odugbemi and Thomas Jacobson, 113–24. Washington, D.C.: World Bank, 2008.

Post, Lori Ann, Amber N. W. Raile, and Eric D. Raile. "Defining Political Will." *P&P: Politics & Policy* 38, no. 4 (August 2010): 653–76.

Prüss-Üstün, Annette, et al. *Preventing Disease Through Healthy Environments: A Global Assessment of the Burden of Disease from Environmental Risks.* Geneva: World Health Organization, March 2019.

Rignot, Eric, et al. "Four Decades of Antarctic Ice Sheet Mass Balance from 1979–2017." *Proceedings of the National Academy of Sciences* 116, no. 4 (January 22, 2019): 1095–1103.

Ripple, William J., et al. "World Scientists' Warning to Humanity: A Second Notice." *BioScience* 67, no. 12 (December 2017): 1026–28.

Roberts, David. "Discount Rates: A Boring Thing You Should Know About (With Otters!)." *Grist*, September 24, 2012.

Rockström, Johan, et al. "Planetary Boundaries: Exploring the

Safe Operating Space for Humanity." *Ecology and Society* 14, no. 2 (2009).

Sánchez-Bayo, Francisco, and Kris A. G. Wyckhuys. "Worldwide Decline of the Entomofauna: A Review of Its Drivers." *Biological Conservation* 232 (April 2019): 8–27.

Simon, Julian. *The Ultimate Resource 2.* Princeton, NJ: Princeton University Press, 1996.

Stern, Nicholas. *The Economics of Climate Change: The Stern Review.* Cambridge, UK: Cambridge University Press, 2007.

Sterner, Thomas, and U. Martin Persson. "An Even Sterner Review: Introducing Relative Prices into the Discounting Debate." *Review of Environmental Economics and Policy* 2, no. 1 (Winter 2008): 61–76.

Storm, Servaas, and Enno Schröder. "Economic Growth and Carbon Emissions: The Road to 'Hothouse Earth' Is Paved with Good Intentions." Institute for New Economic Thinking, December 26, 2018.

Thompson, Derek. "The New Servant Class." *The Atlantic*, August 12, 2019.

Tol, Richard S. J. "Targets for Global Climate Policy: An Overview." *Journal of Economic Dynamics & Control* 37, no. 5 (May 2013): 911–28.

UN Environment Programme. "Emissions Gap Report." November 27, 2018.

University of Exeter. "Nine Climate Tipping Points Now 'Active,' Warn Scientists." *Science Daily*, November 27, 2019.

University of Exeter. "Scientists Search for Regional Accents in Cod." *Science Daily*, October 6, 2016.

U.S. Global Change Research Program. *The Climate Report: National Climate Assessment—Impacts, Risks, and Adaptation in the United States.* Brooklyn, NY: Melville House, 2019.

Vinik, Danny. "The Hidden Impact of Trump's Energy Executive Order." *Politico*, March 30, 2017.

Vita, Gibran, et al. "Connecting Global Emissions to Fundamental Human Needs and Their Satisfaction." *Environmental Research Letters* 14, no. 1 (January 2019).

Wagner, Gernot, and Martin L. Weitzman. *Climate Shock: The Economic Consequences of a Hotter Planet.* Princeton, NJ: Princeton University Press, 2015.

Warren, R., et al. "The Projected Effect on Insects, Vertebrates, and Plants of Limiting Global Warming to 1.5°C Rather Than 2°C." *Science* 360, no. 6390 (2018).

Weitzman, Martin L. "Fat-Tailed Uncertainty in the Economics of Catastrophic Climate Change." *Review of Environmental Economics and Policy* 5, no. 2 (Summer 2011): 275–92.

———. "Fat Tails and the Social Cost of Carbon." *American Economic Review: Papers & Proceedings* 104, no. 5 (December 7, 2013): 544–46.

Wood, Mary Christina. *Nature's Trust: Environmental Law for a New Ecological Age.* New York: Cambridge University Press, 2014.

World Economic Forum. "The Global Risks Report 2019." January 2019.

INDEX